Science, Computers, and People
From the Tree of Mathematics

Science, Computers, and People
From the Tree of Mathematics

Stanislaw M. Ulam

Mark C. Reynolds, Gian-Carlo Rota
Editors

with a
Preface by Martin Gardner

Birkhäuser
Boston • Basel • Stuttgart

Library of Congress Cataloging in Publication Data

Ulam, Stanislaw M.
Science, computers and people.

Includes index.
1. Mathematics—Addresses, essays, lectures. 2. Science—Addresses, essays,
lectures. 3. Computer—Addresses, essays, lectures. I. Reynolds, Mark C. II. Rota,
Gian-Carlo, 1932-   . III. Title. QA7.U43 1986 510 84-21689 ISBN
0-8176-3276-X

CIP-Kurztitelaufnahme der Deutschen Bibliothek

Ulam, Stanislaw M.:
Science, computers, and people. From the tree
of mathematics / Stanisaw M. Ulam. Mark C.
Reynolds ; Gian-Carlo Rota, ed. – Boston ;
Basel ; Stuttgart : Birkhäuser, 1986.
    ISBN 3-7643-3276-X (Basel . . .)
    ISBN 0-8176-3276-X (Boston . . .)

ISBN 0-8176-3276-X
ISBN 3-7643-3276-X

TABLE OF CONTENTS

PREFACE

STANISLAW MARCIN ULAM, or Stan as his friends called him, was one of those great creative mathematicians whose interests ranged not only over all fields of mathematics, but over the physical and biological sciences as well. Like his good friend "Johnny" von Neumann, and unlike so many of his peers, Ulam is unclassifiable as a pure or applied mathematician. He never ceased to find as much beauty and excitement in the applications of mathematics as in working in those rarefied regions where there is a total unconcern with practical problems.

In his *Adventures of a Mathematician* Ulam recalls playing on an oriental carpet when he was four. The curious patterns fascinated him. When his father smiled, Ulam remembers thinking: "He smiles because he thinks I am childish, but I know these are curious patterns. I know something my father does not know."

The incident goes to the heart of Ulam's genius. He could see quickly, in flashes of brilliant insight, curious patterns that other mathematicians could not see. "I am the type that likes to start new things rather than improve or elaborate," he wrote. "I cannot claim that I know much of the technical material of mathematics. What I may have is a feeling for the gist, or maybe only the gist of the gist."

Of course Ulam was being too modest. He knew a great deal about the technical side of math. But it was seeing the gist, the inner core of a problem, that enabled him to open so many new roads — roads that often led to new branches of mathematics. To mention only three: Cellular automata theory, which he proposed to von Neumann; the Monte Carlo method of solving intractable problems, not only in probability theory but in areas such as number theory where the method would not have been thought applicable; and his work on nonlinear processes that anticipated today's interest in solitons and "chaos." We still do not know — it remains a government secret — what jumped into Ulam's head that made it possible for Edward Teller to build an H-bomb.

To me the most fascinating passages in this collection of what Ulam liked to call his "little notes" are those that touch on deep philosophical mysteries. To what extent are physical laws, the patterns of nature's carpet, "out there," independent of you and me? To what extent are they the free creations of human minds? Why is there such an incredible "fit" between elegant theories such as relativity and quantum mechanics and the way the universe behaves? In chapter 2 Ulam recalls telling his friend Enrico Fermi how astonished he was when, just out of high school, he learned how fantastically accurate is the Schrödinger equation of quantum mechanics in giving spectral lines. "You know Stan," Fermi replied, "it has no business being that good."

But the equations of physics *are* that good. Even more surprising, they are often equations of a very low order. Is this because the patterns of nature are, as Einstein believed, basically simple (whatever simplicity means here), or will the laws of the future be increasingly complicated? "It does not say in the Bible," Fermi once remarked (chapter 2), "that the fundamental equations of physics must be linear."

Some physicists believe they are on the verge of creating a grand unified theory in which a single superforce will explain all known forces and particles. There will then be nothing new for physics to discover, and science will turn to life and the human mind as the only phenomena not yet understood. Ulam did not share this view. Repeatedly he cites new developments in physics, logic, and mathematics — especially the work of Kurt Gödel — to suggest that no finite set of laws will ever embrace all there is.

In Ulam's vision the universe is infinitely mysterious and always full of yet-to-be-discovered wonders. Going down into the micro-world there may be endless structural levels, wheels within wheels, and the same may be true in the other direction. Space and time, on a level smaller than a particle, may be discontinuous — a kind of "foam" (as John Wheeler calls it) of holes within holes, and subject to strange non-Euclidean topologies of which we now have no inkling. Similarly, the universe in the large may be embedded in vaster regions where space and time are utterly unlike the spacetime of classical relativity.

It may be true that any day now physicists will find a way to unify the four forces of nature, perhaps even explain how the explosion that created our universe was an inevitable consequence of fluctuations in a primordial quantum vacuum. But the vacuum of quantum mechanics is a far cry from metaphysical "nothing." Its fluctuations obey rigid quantum laws which must be "there" to sustain the eternal dance of virtual particles that bubble out of the energy sea and quickly die. Will the human mind ever learn all there is to know about nature's laws?

Ulam saw the search as endless, both in science and mathematics. "Just as animals play when they are young," he writes in chapter 13, "in preparation for situations arising later in their lives, it may be that mathematics is to a large extent a collection of games. In this light, it has the same role and may be the only way to change the individual or collective human mind to prepare it for a future that nobody can now imagine." This poetic sense of an open future with mysteries forever unresolved, with new surprises forever turning up, pervades all of Ulam's nontechnical writings.

Gian-Carlo Rota, who coedited this volume, has spoken of a writing style that he calls "Ulamian" — a mix of crystal clear prose, subtle humor, and graceful phrasing. The style was the product of a collaboration. Ulam's French-born wife, Françoise, was what she calls in her introduction her husband's "live word processor." It was this happy collaboration that resulted in Ulam's marvelous autobiography. As Ulam phrased it in his introduction, it was Françoise who "managed to decrease substantially the entropy" of his memoirs. The information, speculation and philosophical insights in the book you now hold are all Ulam's, but for the pleasure you experience in reading its essays you have a remarkable duo to thank.

Martin Gardner

AT THE MEMORIAL SERVICE
FOR S.M. ULAM

THE LODGE, LOS ALAMOS, NEW MEXICO, MAY 17, 1984

STAN ULAM resented being labelled an intellectual. He would not even agree to being classified a mathematician. He referred to the published volume of his scientific papers as "a slim collection of poems."

Throughout his life, his style in speaking and writing remained the aphorism, the lapidary definition, the capture of a law of nature between one subject and one predicate. "Whatever is worth saying, can be stated in fifty words or less," he used to admonish us, and to teach us by his example.

Mathematics is a cruel profession. Solving a mathematical problem is for most mathematicians an arduous and lengthy process which may take years, even a lifetime. The final conquest of the truth comes, if ever, inevitably tinged with disillusion, soured by the realization of the ultimate irrelevance of all intellectual endeavor.

For Stan Ulam, this process took place instantaneously, in a flash, unremittingly, day and night, as a condition of his being in the world. He was condemned to seeing the truth of whatever he saw. His word then became the warning of the prophet, the mumbled riddle of the Sybil in her trance. The comforts of illusion were denied to him.

His eyesight followed the bidding of his mind. He could focus on a detail so small as to have been missed by everyone. He could decipher a distant rumbling that no one could yet be aware of.

But his blindness for average distances kept him from ever enjoying a rest in the quiet lull of mediocrity.

Worried that we might not be ready to bear the burden of those visions, he solicitously improvised daily entertainments, games into which he prodded us all to be players, flimsy amusements and puzzles he built out of his concern that we be spared, as he could not be, the sight of the naked truth.

What saved him, and what was to immensely benefit science, was his instinct for taking the right step at the right time, a step which he invariably took with a scintillating display of elegance.

The inexorable laws of elegant reasoning, which he faithfully observed, became his allies as he drew out the essentials of a new idea, a gem of the mind that he would casually toss off at the world, always at the right time, when ready to be pursued and developed by others. His ideas have blossomed into theories that now grace the world of science. The measurable cardinals have conquered set theory; his foundations of probability have become bedrock. He invented more than one new stochastic process, starting from the imaginary evidence he alone saw beyond the clumsy array of figures spewed out by the very first computers. The strange recurrences of the dynamical systems he was first to describe and simulate are the key to the new dynamics of today.

Stan Ulam came to physics comparatively late in life. With unerring accuracy, he zeroed onto the one indispensable item in the baggage of the physicist, onto the ability to spot and shake the one essential parameter out of a morass of data. In his work at the Lab, he was the virtuoso who outguessed nature, who could compute physical constants old and new to several decimal places, guided only by an uncanny sense for relative orders of magnitude.

Every day at dawn, when most of New Mexico is asleep, Stan Ulam would sit in his study in Santa Fe and write out cryptic outlines on small pieces of paper, often no larger than postage stamps. Rewritten, reformulated, rebroadcast by others to the four corners of the earth, these notes became the problems in mathematics that set the style for a whole period. To generations of mathematicians, Ulam's problems were the door that led them into the new, to the first sweet taste of discovery.

I wish we could have convinced him that his problems will last longer than he expected them to last, that they are and will be the source of much mathematics that is and will be made, that he will still find them around in a next life, sprinkled in the research papers and in the textbooks of whatever time; to convince him that they will brighten our lives, that they will brighten the lives of those who come after us, like a cascade of stars in the crystal sky of Los Alamos, like the fireworks of the Fourth of July.

Gian-Carlo Rota

INTRODUCTION

WHEN THE EDITORS OF THIS VOLUME asked me to write a brief introduction, I wondered what a non-scientist like me could convey to the reader about this collection of essays which range from pure mathematics to natural sciences, computers, space, the Polish School of Mathematics and personalities like Banach, von Neumann and Gamow.

My cue came from Ulam's reminiscences of the Polish school, of their working habits; how they used to gather in a Lwów Café to talk and scribble equations and theorems on the marble topped tables before inscribing them in their now famous Scottish Book.

What then are Stan Ulam's own work habits? How were these essays prepared, composed and written?

Every author has his own writing technique. Von Neumann used to save a part of each day to sit down and write what was on his mind at the moment, in page after page of clear, methodical longhand, no matter where he was or what he was involved in. I saw him leave a party at his house, go to his desk, write for a while and come back undetected to rejoin the fun. The eccentric Hungarian mathematician, Paul Erdös, puts down his thoughts in a multitude of letters which usually begin with some sentence like "Suppose X is…" and only towards the end does he indulge in some personal remark about his next travel plans or the state of the universe.

Not so Ulam, for he is almost exclusively a talking man, a verbal person. When not thinking, which he prefers to do while absently laying out solitaire card games on his bed, what he enjoys most is to talk, to discuss, to argue, to converse, with friends and colleagues. Relying on his phenomenal memory, he carries everything in his head. Or almost,…for his pockets, his desk, his briefcase are filled with scrawny, pythic, undecipherable scribblings written on tiny scraps of paper, which are then folded. Disconnected, unrelated, random, they are incomprehensible for anyone else. Once written, these little fragments just pile up, their contents securely filed in his memory. They have served their mnemonic purpose. I have seldom seen him refer to any of them, except during a lecture. Woe just the same to anyone who suggests throwing them away!

The physical act of taking pen to paper has always been painful for him. His mind and his eyes are the obstacles. His mind, because it works so much faster than his fingers that they cannot follow and begin to wriggle like the stylus of a seismograph-hence the jagged, compressed, angular appearance of his handwriting; his eyes, because one is very myopic, the other very presbyotic. Though he claims that by using them separately his vision is better than normal, it is obvious that he has great difficulty focusing at an intermediate range. From childhood fears, then from youthful vanity he spurned wearing glasses, until very recently. Thus Ulam has always had a very hard time bringing himself to write anything for publication, either in long hand or with a typewriter. Machines and other mechanical objects have always turned him off. He is barely able to find and press the record button on a tape recorder. How then does he ever produce a written text? Mainly by talking.

Discussions and conversations with collaborators led to the production of joint papers. Then dictation became his salvation when he discovered secretaries at the Los Alamos labs. From that time on his output increased; writing reports, papers and essays became less difficult for him. It is still his custom, though, never to prepare or dictate a lecture in advance. He finds it more spontaneous to just jot down a few topic headings on one of his ubiquitous scraps of paper. Only afterwards will he reluctantly consent to revise a taped transcript or dictate for publication an approximation of what he talked about.

When retirement cut him off from personal secretaries, I volunteered to become his live word processor and he let me transcribe, edit, cut and paste to my heart's content, until a presentable text can be put under his eyes. Then, and only then will he look at it and add a few finishing touches, always apt and relevant, till he is satisfied with the substance. The implacable rules of logic and precision of mathematical writing have conditioned him to shun anything that is not absolutely factual or a product of rational judgment. He dislikes hyperbolic imagery and prefers to stay away from subjective or affective descriptions.

Glad to have skipped the intermediate steps, he then punctiliously scrutinizes the form, preferably when the text has reached the proof or galley stage. Since Polish was the language of his youth his sentence structure still tends towards the slavic way of developing, refining, pinpointing thoughts as the phrase develops. A long way from the simpler English mode. But if the editors or I have straightened a sentence to a point where he does not recognize it anymore, he will painstakingly return to the original version.

These in short are some very personal views of the whys and therefores of a few ingredients of what Gian-Carlo Rota calls the "Ulamian" style.

Françoise Ulam

ACKNOWLEDGMENTS

BIRKHÄUSER-BOSTON, INC. gratefully acknowledges the courtesy and cooperation of the many publishers who have granted permission to reprint the essays in this volume.

"The Applicability of Mathematics," reprinted from *The Mathematical Sciences: A Collection of Essays,* M.I.T. Press, 1969, pp. 1–6, by permission of M.I.T. Press.

"Physics for Mathematicians," reprinted from *Physics and Our World: A symposium in honor of V. F. Weisskopf,* American Institute of Physics, 1976, pp. 113–121, by permission of the American Institute of Physics.

"Ideas of Space and Time," reprinted from Rehovot, winter 1972–73, pp. 29–33.

"A First Look at Computing, A Personal Retrospective" reprinted from "A Mathematical Physicist Looks at Computing," Rehovot, vol. 9, no. 1, 1980, pp. 47-50; both by permission of the Weizmann Institute of Science.

"Computers" reprinted from *Scientific American,* September 1964, pp. 203–217 by permission of Scientific American, Inc. and the W. H. Freeman Company, Inc.

"Experiments in Chess on Electronic Computing Machines," reprinted from *Computers and Automation,* Sept. 1957, Berkeley Enterprises Inc., by permission of Berkeley Enterprises, Inc.

"Computations in Parallel," reprinted from "On Some New Possibilities in the Organization and Use of Computing Machines, I," IBM Research Report RC–68, May 1957, by permission of IBM, Inc.

"Patterns of Growth of Figures," reprinted from "On Some Mathematical Problems Connected with Patterns of Growth of

Physics Teachers, by permission of the American Journal of Physics.

"Kazimierz Kuratowski," reprinted from *The Polish Review,* vol. 26, No. 1, 1981, pp. 62–66, by permission of The Polish Institute of Arts and Sciences, Inc.

"Stefan Banach," reprinted from *The Bulletin of the American Mathematical Society,* vol. 52, No. 7, pp. 600–603, 1946, by permission of the American Mathematical Society.

"A Concluding Paean," reprinted from *Selected Studies in Physics, Astrophysics and Mathematics,* S. Rassias, ed., North Holland Physics Publishing, 1982, pp. ix-x, by permission of North Holland Physics Publishing, Inc.

CHAPTER 1

THE APPLICABILITY
OF MATHEMATICS

CURRENT RESEARCH in mathematics tends toward ever-more-varied abstraction. Yet the most far-reaching excursions into mathematical theory may lead to applications not only within mathematics itself but also in physics and the natural sciences in general. While it is true that much of the work published in mathematical journals consists of detailed and rather specialized investigations, one might think of them as "patrols" sent into the unknown in all directions. Some of these turn out to encounter new areas of interest in the great game that the human mind plays with nature. Even though much of the activity in mathematical research may appear to be "getting off on tangents," the great sphere of knowledge is increasing at an important rate, and the applicability of mathematics to problems suggested by new discoveries seems to know no boundaries.

Mathematics as we now know it has historically a twofold origin. In antiquity the beginnings of two modes of mathematical thought arose almost simultaneously and probably independently: the arithmetic one — the wondering about numbers — and the geometric one, concerned with figures. It is perhaps remarkable that on the whole the same persons cultivated both these modes of thought. The idea of examining the thought process itself, that is, becoming conscious of logic and of the mathematical method, probably came somewhat later. A certain dichotomy persists in mathematical work even at present.

1

The mathematical *method,* as presently used, probably would not appear strange to the Greeks. However, the *objects* to which mathematical thought is devoted today have been vastly diversified and generalized. It is their proliferation that would perhaps appear so striking not only to the ancients but even to mathematicians of the last century.

The manifest usefulness of applying the notion of numbers to geometric objects, for example, in ideas of length, area, volume, can serve as a first example of the applicability of mathematics. If we consider the beginnings of geometry as observational work concerning the properties of physical reality rather than as exercises in pure thought, one might say that already in the prehistory of mathematics it is hard to make precise distinctions, epistemologically, between pure and applied mathematics. And yet the different psychological motivation and the different emphasis in method allow us to talk meaningfully of this distinction. Perhaps mathematics is unique among activities of the human mind in being, on the one hand, so much an art for art's sake and providing, on the other, so many tangible applications that change the course of the human condition. But in turn these applications mold to a certain extent the development of this abstract art itself.

Philosophically it is equally curious that mathematical idealizations which at first sight seemed "irrational" have led to the most useful and practical consequences. It is the idea of infinity both "in the large" and in the "infinitesimally small" that is so overwhelmingly successful. *A priori,* it is not obvious why the algorithms of the calculus — compared with the operations with finite differences that are more palpably suggested by the first experiences with numerical quantities — should be so convenient, elegant, and powerful in their use. The many applications of the infinitesimal calculus in the natural sciences, especially in astronomy and physics, are to this day perhaps the greatest triumph of mathematical thought. As for the "infinities in the large," the laws of large numbers in probability theory and statistical mechanics allow for more convenient and penetrating formulations than any system of finite inequalities.

Even in pure mathematics itself (for example, in number theory), the asymptotic theorems reveal regularities and give better insight than the more "local" theorems. The ergodic theorems give equally good insight into the properties of many mathematical and physical phenomena.

It is impossible to conceive of present day technology or, indeed, the development of the exact sciences and the technology of the nineteenth century without the previous invention and availability of the infinitesimal calculus, beginning with its role in celestial mechanics and astronomy and followed by its penetration of all of classical physics and its involvement in most practical engineering achievements. The achievements of electricity and magnetism are equally unthinkable without the expanded apparatus of mathematical analysis which permits continuous fields to be described in an adequate fashion. At the same time, some of the mathematical ideas of the nineteenth century, such as those of the non-Euclidean geometries, found their role and application in physical sciences only in the present century. One knows the importance of the ideas of Riemann in the theory of relativity. The second half of the nineteenth century was especially fertile in purely mathematical constructions that have found and are increasingly finding applications in physics.

In this work we shall try to select somewhat arbitrarily, illustrations of these applications, and we shall attempt to give examples of similar use of mathematical ideas of the twentieth century up to the present time.

The great quantum step in the development of the mathematical outlook made by Cantor's set theory has not only affected profoundly the foundations of mathematics but has enlarged enormously the generality of mathematical objects. The theory of very abstract sets leads in natural fashion to consideration of more general structures to serve as spaces for geometrical studies. It was equally natural to consider more general functions than the ones forming the primary objects of classical analysis, for example, arbitrary continuous functions without the conditions of differentiability or analyticity previously imposed on them. The preponderant study of functions of real or complex variables was enlarged to form a study of transformations of spaces. It was equally natural to consider classes of functions as themselves forming spaces, endowed with geometries of their own. Functional analysis became an impressive edifice of mathematical work.

The generality of the idea of space and the successful employment of these notions stimulated a parallel development of very general abstract algebraical structures. All this very extensive work can be regarded as an application of new mathematical abstractions to the mathematical objects previously considered. One striking example is the use of fixed-point theorems, generalized from finite dimensional Euclidean spaces to spaces of functions. These are important in establishing the existence of solutions of differential or integral equations. As another example, we may mention ergodic theorems. These theorems, proved in a framework of the theory of measure of sets and of real variable theory, lead to theorems like the laws of large numbers in the theory of probability. One could give numerous other examples of the applicability of the more general mathematical methods to older, more "classical" problems. The more abstract and purely combinatorial thinking was found to be of great use in problems of classical number theory. This is evidenced, for example, by the work of Schnirelman on the problem of representing all integers as sums of a fixed number of primes.

The theory of groups developed from the study of purely algebraic problems by Abel, Galois, and their successors. In the hands of F. Klein, this development found a programmatic role in the foundations and planning of geometric theories. But in the course of the last few decades the theory of groups has begun to play an increasingly important role in the very foundations of physical theories. It has provided most important tools for the ordering and the classification of atomic spectra. It is through group theory that the apparent chaos of the spectral lines can be replaced by an order deriving from general principles of quantum theory. More fundamentally important still, starting at the beginning of this century the special theory of relativity was able to impose its all-embracing role through use of the notion of the group of transformations under which physical laws have to be invariant. The important thing to remember is that not only the transformations of the Minkowski spaces in themselves but the abstract properties of the group which they constitute were essential for this purpose. The Lorentz group is thus one of the most important ideas in all of mathematical physics.

In more recent times, the notions of group theory have become very useful and perhaps crucial for the understanding of the variety of "fundamental" particles. We cannot describe the origin and the scope of these ideas in this work, but shall have to content ourselves with mentioning that at the present time important work is proceeding to account for the variety and properties of the elementary particles through assumptions of the existence of a few groups of symmetries governing, as it were, the choices of nature. In addition to the Lorentz group, these symmetries seem to involve certain more finitely describable ones, for example, symmetries of the spin, the charge, and the duality between particles and antiparticles.

It is quite well known that the development and the use of mathematical tools were a necessary prerequisite, and indeed often a stimulus, to the development of present day technology. The language and the technique of calculus permitted not only the developments of the machine age, such as the construction of bridges and the design of electric motors, but also the formulation of theories of thermodynamics, electric fields, chemical reactions, flight in the atmosphere, and so on. The recent achievements in rocket propulsion and the construction of artificial satellites are all based on principles of classical mechanics, a vast field of application of mathematical analysis to consequences of Newton's laws. So much of all this is taken for granted that we find it worth while to remind ourselves of this obvious and common role of mathematics.

Perhaps not so well known is the role of mathematical thinking and mathematical techniques in some other technological achievements of this century. It should be quite obvious that the atomic era could come only after the great discoveries of theoretical physics — of the equivalence of mass and energy, which followed from the theory of relativity — and the better understanding of properties of matter in general, which came through quantum theory. For the construction and technology of the atomic reactors, the apparatus of mathematical physics was essential. In the construction of atomic bombs and of the hydrogen bomb, an enormous amount of mathematical work had to be done. This involved not only the techniques of classical analysis but also some of the more recent mathematical developments that made possible modern thermodynamics: statistical mechanics of material particles and of fields of radiation.

The theory of probability pursued by mathematicians in this century and the theory of spaces with infinitely many dimensions, like Hilbert space, have found applications. In the work now proceeding on the construction of fusion reactors (that is, the attempt to design devices which, in contrast to the H-bomb, would release the energy gradually), an enormous amount of mathematics of the most sophisticated type is being used. It is very interesting to some "pure" mathematicians to see how some very abstract-looking theorems (for example, on the existence of fixed points of transformations and of periodic orbits) find important applications in the design of the big accelerating machines propelling protons or electrons to velocities very close to that of light. The increasing complexity of not only the problems posed by technology but the formulations of the very foundations of physics makes an increasing demand on the use of the most modern and complex mathematical ideas. The problems to which the most advanced mathematical ideas are being applied include such ambitious enterprises as that of calculating the general circulation of the atmosphere of our globe and of making weather predictions. Mathematically speaking, this involves systems of partial differential equations in three spatial dimensions and in the time variable. Mathematical work has been greatly expanded and is now proceeding on the understanding of the qualitative and quantitative behavior of the solutions of such systems.

In very recent times another area of the natural sciences appears to be becoming mathematizable. This is the great field of molecular biology. The recent fundamental discoveries in this field begin to form a framework for theories promising an understanding of the life processes. The fundamental processes of replication of a living cell, the code describing the construction of materials necessary for what we call life and its fundamental processes, begin to emerge as a rationally describable set of schemata. It is hoped that the ideas of structures of mathematical systems in general and especially the field of mathematics known as combinatorial analysis will be useful in working out the perplexing complexity of these arrangements.

The future understanding of the function of the nervous system in living organisms and some of the mysteries involved in the workings of the human brain itself may be gradually aided by new mathematical ideas and techniques.

In all these new problems, presently under full study, a new tool has been and will increasingly be of decisive use: electronic computing machines. As we mentioned before, the new and more ambitious problems to which mathematics is applied are characterized by a complexity far exceeding that encountered in previous problems of physics and technology. The speed with which computers operate allows performance of the billions of operations necessary to computation of many such complex phenomena.

The recent achievements in space technology would be unthinkable without the availability of both telemetry and electronic computers which together permit almost instantaneous calculations of the changes in propulsion necessary for the achievement of prescribed orbits.

In the calculation of the motion of air masses for weather forecasting computers are not only useful but absolutely necessary.

Beyond all this, the designing and the operation of electronic computers themselves involve ideas of mathematical logic and combinatorial analysis which are themselves developments of the present century and of the present time.

Studies of these methods also promise to be of use in the study of the functioning of the nervous sytem and of the human brain.

CHAPTER 2

PHYSICS FOR MATHEMATICIANS

WHEN I FIRST CONTEMPLATED writing about physics I had great hesitation. Then it occurred to me that if Viki Weisskopf can conduct a symphony orchestra, maybe I can write about physics!

This chapter *PHYSICS FOR MATHEMATICIANS* will almost mean physics without mathematics. My interest is, to paraphrase a famous statement, not what mathematics can do for physics, but what physics can do for mathematics. That is my underlying motive.

In the last decades, there has been a widening gap between theoretical physicists or people who know physics, and the professional mathematicians. Perhaps this is unavoidable because within these sciences themselves there is increased specialization. However, in the nineteenth century there were mathematicians who were also physicists or who contributed very significantly to physics. For example, at the end of the last and beginning of this century there was Poincaré whose marvelous popular books on science in general and physics and mathematics in particular had such enormous influence on the young people who read them and still could see the inspiration both subjects derived from each other. Besides, Viki's article on the impact of quantum theory on modern physics begins with a quote from Poincaré. The lines he quotes were written about the time Viki was born (which was not so terribly long ago). If you read a Poincaré book, something I would recommend even now for historical perspective, you will see how much has happened in this period of time.

What I want to discuss now are the conceptual possibilities which now more than ever, the historical spectacle of physics provides to mathematicians, even the most abstract ones. At the same time I will mention a few mathematical ideas which will perhaps play an important role in the physics of the future, near future, or even more remote future. But, of course, that is always difficult. As Niels Bohr said, it is very hard to predict, especially the future. Poincaré was one of the last universal minds of this sort. In more recent times, one could mention Herman Weyl and to some extent, John von Neumann. They contributed more than mere grammar to the science of physics; their points of view influenced the physical thinking itself. I will describe several areas of what we call physics which might stimulate mathematicians, together with some mathematical thoughts which have influenced at least the techniques of physics.

One such field known to perhaps everybody is group theory. At first, one might have thought group theory merely played the role of a very good filing system, but it has assumed an important fundamental role. Here I come back to Poincaré's books. There are four books; they are very popular and are on the very highest level, both from the point of view of literary merit and philosophical merit. They are called *Science and Hypothesis, Science and Method* (the one I want to refer to), *The Value of Science* and *Last Thoughts*. The one I wish to refer to was written around 1908. It contains some of his speeches at the World's Fair in St. Louis in 1904. Among them is a very famous address on the unsolved problems of mathematics and physics where he discusses the discovery of certain harmonies of nature as one of the aims or one of the most important stimulations for work in physics. In this word "harmony," he points out, are more general ideas than the idea of a group. It embodies all of the analogies, formal, obvious or hidden in various parts of physics. To discuss harmony Poincaré used the, by now so familiar, role of linear partial differential equations in several apparently quite separate and different parts of physics. It is really very strange to look at this book now and compare the picture which one had of the world of physics, of nature, with what is now known. And there are some curious little items. I noticed he mentions the Japanese physicist, Nambuma, who around 1900 or a little before, certainly before Niels Bohr, proposed that the atom is one big positive electron surrounded by a ring of negative electrons and draws the conclusions which, of course, later led to Bohr's model.

The ideas of Poincaré had a different impact on the development of mathematics from those which Hilbert formulated in his famous lecture to the Paris Congress around 1900. Indeed Poincaré himself at the same congress gave a lecture on the different mental attitudes within mathematics itself. He said there are two types of mathematicians. There are the analytic ones who given some axioms or rules like to produce new ones with utmost ingenuity, or like to dissect and construct new objects of mathematical thought. And there is the other type, the intuitionists, who, without being so much interested in the formal aspect, divine by analogies the pattern of the world of thought. Clearly such distinction must exist among physicists. There are people who use as virtuosos, the mathematical techniques; and there are others who stress more the concepts in physics. It is this intuitive type which is *prima facie* useful to mathematicians who are curious about the physical world.

The situation in mathematics has led to increased fragmentation and specialization. By now there are ten or twelve different kinds of topological methods or fields and often the practitioners of one do not know much about another.

Once I was asked to give a talk at some jubilee or memorial; actually it was the 25th or 30th anniversary of the Princeton computer. As I was talking in general terms about mathematical problems which can be studied by heuristic experimental work on computers, it occurred to me to estimate the number of theorems which are proved yearly. Defining theorem as something called "theorem" which is published in a recognized mathematical journal, I quickly multiplied the number of journals by the number of issues by the number of papers in each issue and by the average number of theorems. When I got the answer I thought I had made a mistake. I said one hundred thousand theorems are proved yearly and the audience gasped and I was somewhat appalled and thought I must have made a mistake of an order of magnitude. But the next day two younger mathematicians who were in the audience came to me and said they had made a better search in the library and they found out that the number was closer to two hundred thousand yearly. Will that sort of thing happen in physics? Will there be again proliferation and fragmentation? There is much less danger of it. There is a certain unity in physics and even physicists who specialize in this or that field are aware of others. Of course, in physics there is a proliferation of publications, too. Somebody told me that if one extrapolates the number of pages in the *Physical Review,* there would be so many that even at light velocity you wouldn't be able to peruse them all. Physics, however, still keeps its certain unity. Physicists, no matter how specialized they are, on the whole have interest in and knowledge of the foundations of physics which is not at all the case in mathematics.

Now, to be more specific, I would say one problem that would interest a mathematician, one who is familiar with the idea of physics, would be the problem physicists do not consider a physical problem. The problem is this: Is the world fully describable by physical laws? Is it finite or infinite? I mean it in the following sense. Historically there was always, since the Greeks, the idea of atoms which are indivisible and unanalyzable. During the last, let's say sixty years, the molecules were found to be composed of atoms, atoms in turn of nuclei and electrons. The nucleus itself was a group of nucleons and now, as you know, people take a little bit more seriously the picture of a single nucleon being composed of partons, perhaps quarks, perhaps something else. The mathematician's question is: Will this go on, perhaps forever? It is a question which, until recently, you could have said is merely philosophical. But perhaps less so now. If the things are very small and in a certain sense, throw a shadow ahead of them, perhaps it might be possible sometime in the future to consider the world infinite in structures that go on through stages which are not at all necessarily similar to each other. The same question was well considered long ago in the large: Is the universe of stars, galaxies and the groups of galaxies bounded? Is the metric elliptical or is it hyperbolical or is the universe truly infinite? And this to some extent can be investigated or guessed. Perhaps it is easier in both cases to assume the universe is infinite. These are the models which would delight the pure mathematician. Mathematicians like infinities. Physicists are beginning to wonder whether the sub-division of structure will continue forever. I will say this is, as yet, not a problem which one can precisely investigate or even devise experimental tests for. It is more convenient to work with the calculus of infinitesimals than with finite differences in pure mathematics. Is there an analogous situation visible in the next ten or twenty years in theoretical physics? If so, then, a very interesting and, to some people, ominous possibility will appear. As most of you know, in mathematics in the last fifty years, starting with the discoveries of the logician Gödel, it was found that any finite system of axioms or even countably infinite systems of axioms — if not trivial — will allow one to formulate meaningful statements within the system which are undecidable, that is to say, within the system one will not be able to prove or disprove the truth of this statement. Fortunately, I think one is very far from that yet in physics, but it is permitted to speculate whether the number of essentially new phenomena, the number of laws will increase forever, or whether as was the great hope of the 19th century and even of the first part of this century, there will be a few ultimate laws which will allow one to explain the external world.

Psychologically, there is a difference between mathematicians and physicists, but I think it might attenuate in the following sense: Mathematicians start with axioms whose validity they don't question. You might say it is just a game — "the great game" as Hilbert called it — which we play according to certain rules, starting with statements which we cannot analyze further. Now to some extent, at least, I would claim that in physics the situation is inverse. Given a lot of facts, let's call them theorems, we look for axioms, that is to say physical laws, from which they would follow. So physics is an inverse process. And in mathematics itself, you could think of such a game: Given some theorems in a certain well-defined notation or algebra, find the underlying axioms. This game has not been played and I'll tell you why. It is because the idea of algorithm and the formalization of what we call definite equational theory is very new. So mathematicians have never hesitated to create objects of their own. Right now there begins to appear some work which I would consider of this inverse type; perhaps it was stimulated by the existence of computers. The questions for a mathematician is how to approach this vast spectacle of physics, whether it is really a game starting with given laws or whether on the contrary, as seems to be the case, given an enormous number of facts or classes of facts it is to discover the rules from which they follow. It is, at least some physicists think so, the question of formal beauty. I would rather not subscribe completely to what Dirac said several times — first write beautiful equations and then if they are really very nice discover that they are probably correct. He may just have stated this extreme point of view in order to encourage this kind of work. He himself had one of his greatest successes that way. This should interest mathematicians who are logic or set theory oriented. Let me tell you, by the way, that one could say that even the foundations of mathematics itself are written on sand. The recent discoveries in set theory show that the intuitions which were so commonly shared among mathematicians about the degrees of infinity are dependent on more assumptions than we had imagined and to some extent vary from person to person.

Physics, however, is a world in the singular. In mathematics there is not just one geometry, as there was in the beginning with Euclid, there are many geometries. Just as there are many geometries, who knows, there might be several different types of physics. Not merely different in formulation, as the same thing in different languages, but given for instance different situations, (call it other universes), the laws could be different. Not only the constants could change in time but conceivably just as a mathematician's plaything, it could be that some of the rules or laws themselves are variable. This could be stimulating work for mathematicians interested in the world of foundations. My own taste would not run to that at all because it is a *post mortem* activity. It is again a question of taste.

I come to my next topic: What is the stage on which physics is played? Well, it is the structure of space and time. To me, it is rather surprising that the idea of space — Euclidean or Newtonian space — was generalized by mathematicians in literally thousands of papers and books, to very general topologies, very general metrics, while very little was done in a similar vein with the four-dimensional non-positive definite metric of the Lorentz space. There are hardly any speculations on how to generalize space-time and play with it mathematically. The classical ideas of Riemannian surfaces are only very, very special things. Nowadays, mathematicians consider strange topologies in the small which are non-Euclidean even locally, and one day again, it wouldn't do any harm if somebody could write an article on this, similar to the popular lectures of Poincaré even if it is much more difficult. It could stimulate or indicate the possibility of using, at least formally, the idea of a space full of holes within holes. Such things were considered long ago by the creators of set theory and topology. In other words I would say that geometry itself is the stage on which physics is played. I have myself written several "propaganda" articles on these possibilities.

I was fresh out of high school when I read about the new quantum theory. It surprised me that the ψ-function was defined for all points xyz and that xyz were defined with infinite precision. Of course, the difficulties with alternative possibilities are well known. There is no minimal distance possible which would be consistent with the Lorentz special relativity transformation. But it need not be that there is a minimal distance. Just like there are sets between points of which the distances are arbitrarily small but not all of which "exist," only some of the points "really" exist, the others being "virtual."

I have myself written, together with some other mathematicians, little notes on "p-adic" time-space in which other groups would play the role of the Lorentz group. What would be the equivalent of the light cone, etc.? These are sets of points which have the "time" norm equal to the "space" norm. Indeed I have often been struck by how much the classical ideas help in the realm of phenomena where the quantum theory was supposed to be complete and at variance with any sort of mechanical model interpretation. This is the inverse correspondence principle — the classical ideas seem to have validity, at least as models of stimulation, for the very small. Of course, it is very hard to make this transition to the macroscopic phenomena, to sew it together without the seams showing. That is the problem.

I remember Fermi's frequent visits to Los Alamos. On walks I would tell him the impressions of a young mathematician about physics and how the derivations weren't satisfying for some people. I told him that when I first learned, just out of high school, about the success of the Schrödinger equation, and how pulled out from thin air, it gives the spectral line values with accuracy to three or more decimals I would have considered it a fantastic success if it was correct to 10 percent. Fermi said, "You know, Stan, it has no business being that good."

Sometimes in discussions about some problem he would tell me, "Oh, now they say that one should make the following assumption." I was very much amused because, "they," of course, included Fermi, himself, one of the creators of the quantum theory.

Here is a problem related to the nature of space-time, and the well known difficulties of divergences in field theories. The situation is even more interesting than I have so far portrayed, namely, one can ask, "What is the nature of what mathematicians call the primary variable used in physics?" The Newtonian point of view was essentially that points are the primary variables. The idea of the field came with the role of the continuum of points, and is still incredibly successful.

But, think of the Fermi-Dirac and Bose-Einstein statistics

which deal with points but with indistinguishable points. That is really something new again. The ψ-function, the primary variable of quantum theory, is no longer what we consider a geometrical point. A set of points, or a set of sets, is different from the classical collection of points considered in analytic geometry or in most courses in mathematics, even topology.

In 1931 when I was still in Poland, my friend Karol Borsuk and I considered entities which are finite sets of points and tried to make a topological space of these. That is, consider a finite set of points on an interval and do not distinguish between their order. The element is a set of points and the distance between two sets can be defined in a simple way as described by the mathematician Hausdorff. Given a point x of set B, look at the nearest point y in set A and then take the maximum of the minima with respect to all choices of x. That will be the distance between sets. We wrote a paper, the first I ever published in this country. We sent it to the Bulletin of the American Mathematical Society in 1932. In it we showed that for two dimensions you get a space that you might imagine just as a triangle, for three dimensions you can imagine it as a sort of tetrahedron, but topologically they are the same as a square or a cube. For four dimensions it is still topologically the same. And now surprise! For five or more dimensions one gets a strange sort of manifold topologically different from the n-dimensional cube. At that time I had no idea of the indistinguishability of particles, the statistical mechanics theories of Einstein and Fermi. This I merely mention to show that in mathematics, one considers as variable not only points but sets. Sets for a space with respect to distance. As you go to infinitely many dimensions, you get something very interesting in the space of the closed sets. Here again is something that most mathematicians should learn about, namely the statistics of identical particles. Of course these statistics are very unintuitive. Gibbs, himself, found a paradox by considering the entropy of two gases mixing. The entropy increases because of diffusion from one part of the gas to the other, but when the two gases are the same, the formula is different because the diffusion does not count.

This was the precursor of the idea of indistinguishability. We don't know which particle is which and the actual formula for the entropy is different. This difference was of fundamental importance many years later. Here again is a very interesting field of work provided one explains to mathematicians what physics deals with, and what is actually important conceptually.

I would now like to describe a very nice new mathematical game on nonlinear problems. It is the humble and simple beginning of a topic which now has many books and dozens of mathematical papers. The reason I mention it is because no less a physicist than Fermi was involved in it. It was in the very early days of computing and Los Alamos had one of the first working computing machines. I remember discussions in which we looked for good physical problems to study on a computer — a problem in which one could not gauge even the qualitative behavior by calculations as they would take thousands of years. After some time Fermi said, "Well, a nice problem would be to see how a string vibrates if you add to the usual Hooke's law a very small nonlinear force." So this was the problem. We were interested in the rate at which the motion becomes ergodic or mixed up. Fermi, himself, expected that after a while, after many of what would be linear vibrations back and forth, the string would distort more and more. But nothing of the sort happened. To the great surprise of Fermi, who after all had tremendous experience in problems involving wave motion, it didn't want to do that. Only the first few modes played among themselves. The higher modes, beginning with say the fifth and up to the 64th, involved all together only one or two percent of the total energy. So the motion was very nonergodic and in addition it was very surprising that after a few thousands of what would have been the normal period, the string came back almost exactly to its initial position. According to a theorem of Poincaré, the return theorem, this should happen sometime, but the time interval for it would be on the order of the age of the universe if you estimate it by phase-space volume. But no, this happened quickly. Our computer work gave rise later to many studies trying to explain this phenomenon. This nonergodic result is also physically observed in some propagations of sound waves in crystals and by now people know that certain nonlinear problems lead to quasi-states or quasi-eigenvalues. In a linear problem, these characteristic states appear just like in the theory of Hilbert space used in quantum theory. The suspicion was that nonlinear problems led to a general increase of entropy, a mixup of everything. But no, in a large class of problems, even in some problems of classical dynamics, there is no apparent ergodicity under certain conditions.

Fermi said, "Well, it does not say in the Bible that the fundamental equations of physics must be linear." And the idea of quasi-states or what some mathematicians now call solitons, is now being used in some speculations about the more fundamental physical problems, in particular models for elementary particles. Here, again, something that started with very humble, simple-minded calculations has led to interesting mathematical developments and speculations.

In subsequent chapters I shall explore some of these ideas more fully.

CHAPTER 3

IDEAS OF SPACE
AND SPACE-TIME

WE LIVE AT A TIME when the old nineteenth century hopes of achieving a simple description of nature have been disappointed. The complexity and strangeness of astronomical discoveries, of most peculiar stars and collections of stars, the great variety of unexpected extreme conditions, the finding of unexplained luminosities of galaxies, indeed the whole shape and behavior of the universe in the large, present problems of great perplexity.

The world now seems much more mysterious than nineteenth century ideas could possibly have indicated. The observations which give us these indications of its shape, and perhaps also of its past history and future, are now made not only visually, as before, through optical telescopes, but also with the aid of radiation beyond the visual range. These show us "the invisible constituents" of the universe surrounding our solar system, the Milky Way — that galaxy of stars of which our sun is one member — and beyond it the vast assembly of other galaxies. The interstellar space is filled with particles, some extremely energetic — the cosmic rays — and also with an extremely tenuous assembly of individual nuclei, and with a pervading radiation whose properties are perhaps a faint echo of the conditions at the beginning of time!

What is the space and the time of which these objects — the same stars that the ancient peoples of Babylon and Egypt gazed upon — are part? What are the positions and times of the faint luminous signals observed through our telescopes and the oscillations of the waves detected through our instruments? And at the other end of the size scale, going down to the dimensions of molecules, atoms, nuclei, and even the constituents of the nuclei, what is the meaning of size, the meaning of relative positions, of "spatial arrangement" and indeed of the space itself in which these objects are located?

I shall try to sketch, as briefly and simply as possible, some salient points of the history of ideas about space and space-time, and how they have evolved, and touch upon some unresolved fundamental problems.

In every book on the history of science, and on mathematics in particular, we read of the origin of the first ideas of geometry, how they came about both from very practical problems of measurement, such as areas of ground or volumes of goods, and also from the ancient people's observations of the clear night skies of the Near East.

The next great advance came when Euclid developed an **axiomatic** treatment of geometry, that is the science of the descriptions of relations between objects in space. What is important for us here is that Euclid provided an example of an axiomatic system for a vast assembly of abstract "facts", i.e., theorems regarding the properties of objects. At the same time, Euclidean space, as it is now called, was supposed to be a basis for properties of the "real" or physical world in which we live and to describe the stage for the universe of material objects. These objects were considered as either "points" or collections of points, lines, planes, single figures, and so on.

Euclid's monumental work was to provide for centuries both the foundations of geometry and a didactic Canon from which generations learned mathematics. Indeed it reigned supreme until, in the 19th century, Lobachevsky, Bolyai, and Gauss showed that there could exist other systems of geometry, and that the Euclidean one was not the only one possible.

These mathematical constructions did not by themselves challenge the sole validity of Euclidean geometry as a description of the physical space of our universe. It was Riemann who suggested and elaborated an entire class of possible geometries, with a prophetic intuition that they might describe physical reality and perhaps be applicable as models of the universe.

Riemann constructed theoretical spaces where locally (that is, for small neighborhoods of each point) the geometry indeed satisfied the properties of the Euclidean space. However, in the large, globally, these spaces cease to be "infinite and flat," and their overall topological characteristic could be that of a sphere or of a ring surface, or perhaps some other surface with many holes. They could be closed, i.e. bounded or compact, as mathematicians call it, or extend to infinity. He suggested that the spaces of physics might be found among these more general spaces which differed from Euclidean space in the large, though, as we have emphasized, their "local" properties remained Euclidean. However, it was left to the physicist to establish definitions or rules for measuring distances between points of physical space, and to check whether such consequences of Euclidean geometry as the Pythagorean theorem, and the properties of angles, were really satisfied.

Meanwhile figures still more general, each of which could be considered by itself a "space," were becoming the object of mathematical definition and study. The great expansion in the scope of mathematical thought which came about through the development by George Cantor of abstract set theory, led to the study of "arbitrary" sets of points on the real line and in the plane, and with it, later, through the imagination of mathematicians, to an axiomatic definition of general spaces.

One way to define a general space is to imagine just an abstract set of elements which are to be thought of as points, with an assumed idea of which sequences of these elements converge, and to which limit points. This class of sequences can be defined quite arbitrarily provided it satisfies certain simple properties embodying our intuition of convergence, i.e. of increasing proximity. For example, one postulates that any subsequence of a converging sequence converges to the same point.

It may be surprising to a non-mathematician how, from a few

properties of this kind, one can build up an interesting and rich study of such spaces. Alternatively, one can approach the idea of a space still differently, through the notion of a distance between any two elements of an abstract set E.

Suppose again a set E of elements is given in which we now have attached to any two of them, x and y, a real number which will be called the distance between them, and denote it by $\rho(x,y)$. This number should satisfy the following properties: 1) $\rho(x,y) \geq 0$; $\rho(x,y) = 0 \equiv x = y$. This means that the distance is not negative, and is zero if and only if the two points x and y are identical. 2) $\rho(x,y) = \rho(y,x)$, i.e. that the distance from x to y is the same as that from y to x. 3) $\rho(x,z) \leq \rho(x,z) + \rho(y,z)$, i.e. that any side of a triangle is always less than the sum of the other two.

Sets, in which a distance with these properties is defined, are called **metric spaces.** They form a most important class of topological spaces and include, of course, the Euclidean space, the kind of spaces considered above by Riemann, etc. Such metric spaces can be very general.

One such general metric space is that known as the Hilbert space. We shall omit any technical description of it, but should mention that it arose originally in connection with problems of solution of an infinity of equations, and its definition was suggested through problems in algebra and in analysis, not by geometrical considerations. However, developments in theoretical physics led to this Hilbert space in a different way, and it was vital for the foundations of the physical quantum theory that the necessary mathematics had already been worked out and was ready for use.

Still another example of this kind of parallel and independent stimuli is furnished by the consideration of the classical Riemann spaces in the general theory of relativity. Here, too, the mathematical tools were there for Einstein to use for his formulation of the geometry of space. Einstein's theory of special relativity initiated a revolutionary departure from the restrictions on the scope of geometrical ideas by involving, in addition to the physical "space," the variable or the parameter of time.

Following Newton, geometers as well as astronomers and

physicists had considered time as "flowing uniformly," as an independent parameter in the course of which physical changes took place in a static and immovable space. These changes consist of motions of material points or bodies in it or else motions of a continuous "ether" filling it and moving like a fluid, the wave motions describing radiation and electromagnetic phenomena in general. A most radical change in this schema came when Einstein, with bold physical insight, built a union of the three variables of our accustomed space with time as a fourth variable into one four-dimensional continuum, the "space-time" of special relativity theory.

The mathematician Minkowski realized the overwhelming impact of this new schema for theories of the physical world. Shortly after the appearance of Einstein's paper he stated at a congress in Germany: "From now on space itself and time by itself shrink to mere shadows, and only the combination and union of the two together possess physical meaning."

The three space variables, x, y, z, are the coordinates of our Euclidean space; t denotes the Euclidean time parameter. The motions of a point (x, y, t) were classically described by giving $x(t)$, $y(t)$, $z(t)$ as functions of time. The changes in coordinates which leave invariant the physical properties of space include translations and rotations. The Euclidean distance between points is an invariant of such transformations, that is to say, if we change by rotation and/or translation of the coordinate axes x, y, z into x', y', z', the distance remains invariant.

$$(x^2) + (y^2) + (z^2) = (x')^2 + (y')^2 + (z')^2$$

Michelson's experiments which established that the velocity of light was constant and independent of the velocity of the observer led to the realization that physical phenomena are invariant under a different class of transformations in the four-dimensional space of the coordinates, x, y, z, t. It is a different distance that is preserved or is invariant. The distance in question is derivable from the expression: $\sqrt{x^2 + y^2 + z^2 - c^2 t^2}$ (c is a constant — velocity of light) (note the minus sign in front of the variable t).

25

We might say here, by the way, that mathematicians, in contrast to the enormously elaborate work on metric spaces, generalizing, as it were, Newtonian space, have not made anything like the same effort to define general set-ups for "relativistic" metrics, nor any axiomatic theory which would present analogues of space-time, including analogues or generalizations of time-like variables.

This is perhaps curious epistemologically, and I, for one believe that such studies may be soon undertaken by mathematicians.

In relativity theories the merely "spatial" part of space-time does not have any unique or absolute meaning since the distances depend on the velocities with which the "observer's" system of coordinates moves. Thus indeed the geometry of the space part by itself is "relative." Similarly **time** by itself may be dilated as viewed from another moving system of coordinates. The invariant and more "absolute" properties belong only to a larger system, the four-dimensional interval, i.e. distance, which is the Lorentz-Einstein-Minkowski space-time.

To jump a little ahead of our story, let us anticipate the further generalizations used in physics and state some of the logical set-ups of quantum theory, the branch of physics which enables us to deal with particles on the atomic and sub-atomic scale.

Here the fundamental or "primitive" notion is not material mass points, but rather probability distributions, the famous ψ functions which, for our simplified purpose, may be crudely thought of as follows.

Entities like an electron or an atom with a system of electrons around its nucleus are to be thought of as smeared out distributions of probability for the "particles" being at a given place in space at given times, and ψ (x, y, z, t) defines this probability. To be exact it is a function defined on the real space of positions (x, y, z); it has complex values and it is its square which represents the probability of what we habitually visualize as the particles occupying given positions at a given time.

In the same way the velocity of the particle has meaning only as the average of a probability distribution in the variables of components of velocities, or momenta. This is the starting picture in quantum theory which to begin with was still not relativistic. In the original theories of De Broglie, Schrödinger, Heisenberg, time was still featured as an independent parameter.

The first steps to imbue the relativity theory point of view into the quantum theory of probability distributions were made by Dirac, whose technical formulation we cannot easily describe here, even in general terms.

The relativistic quantum theory does not, however, present a completed edifice. What we are trying to stress is that the "space," i.e. the external characteristics of what we are accustomed to consider as particles, are now becoming much more abstract.

For example, the intrinsic or internal "motions" (in the more classical picture to which we are accustomed from macroscopic impressions of our senses we would say "internal" rather than intrinsic), are treated as variables which are called spins. These somehow reflect a rotation inherent in the particle, a rotation around itself would be an intuitive image. But its magnitudes can take only a discrete set of values, and these are independent of the orientation of the axis in space: in other words, the spin is a sort of abstract pointer tied to the electron itself.

The various species of particles have by now become very numerous through discoveries in high energy physics. There is a great variety of "elementary" particles whose characteristics involve several types of abstract spins. All these could really form part of what we should properly call a general space or stage for physical phenomena.

The collisions between such particles give rise to transmutations, and much of present day physical theory concerns itself with attempts to discover and classify properties of what govern the processes leading to such transformations.

At the present time, from a mathematician's purist point of view, the physical theories are far from complete. The nature of what was called ether in the 19th century, or in present day parlance a field in the vacuum, is still not understood, i.e. a satisfactory set of axioms has not yet been found which can describe its behavior.

The foundations of the science called quantum electro-dynamics still present serious and fundamental difficulties, despite the great successes obtained by its formal use, in enabling us to calculate with unbelievable precision subtle phenomena governing minute observations.

What a fertile field for mathematicians who, in contrast to physicists, need not be bound by consideration of the one and only "real" universe, but may entertain visions of many diverse possibilities! Just as 19th century geometers were not bound to what was considered the "true" Euclidean space, but dared to think of more general ones, so now a more general "physics" could be imagined, currently fictitious but perhaps useful in the future.

Let us indeed visualize some possible further generalizations of the ideas of space and time.

We mentioned at the beginning of this article the essential nature of Riemannian spaces in the small, namely their "locally Euclidean" character.

For one-dimensional space the locally Euclidean character means that the neighborhood of a point closely resembles an interval. Ever since George Cantor's work on general sets, mathematicians have considered other sets of real numbers with locally different properties.

An important class of sets of real numbers is one called perfect nowhere dense sets. Here is an example of such a set.

Imagine the interval (0,1) and remove from it the middle one-third (but leaving the end-point of the removed interval). On the two intervals thus left, repeat the same operation to scale, that is erase from each its middle third. We are now left with four intervals (with their end-points) and we can repeat this operation indefinitely. A set of points will remain, which, oddly enough, still has as many points as the whole interval, and looks like an indefinitely continuous Swiss cheese in one dimension. A two-dimensional version of it would be like the surface of the moon, with holes of diminishing sizes appearing in every area.

Analogously, a three-dimensional or n-dimensional analogue may be defined. This set is called a Cantor discontinuum after its discoverer. It may indeed provide a geometrical schema for space-like descriptions of physical phenomena in the subnuclear dimensions. Its elements can be combined together by algebraic operations, and it is not impossible that the substratum of the ψ functions in quantum theory, i.e. the range over which probability distributions are defined, may be of that nature.

What we are speculating is that even though no smallest distance can exist in nature, the meaningful distances may possibly form an infinite scale, but arranged discretely and not necessarily as a continuum.

Indeed, historically, what was at any stage thought of as a continuous, primary and indivisible elementary ball turned out later to be a structure. This process has been repeated several times in recent decades: a molecule turned out to be a system of several atoms, each of which in turn, far from being an indivisible or impenetrable ball, appeared as a small central nucleus surrounded by electrons; the nucleus itself is composed of separate nucleons, neutrons and protons, each of which might be again a system of "quarks" — and perhaps "and so on"?

The mathematics of such hierarchic universes could very well involve a picture of space in the small having a nature analogous to a Cantor discontinuum.

Still another illustration of how the ideas of what we call space may be radically different for physical situations is suggested by the principle of "indistinguishability." As an example, suppose we have three particles, each characterized by a position on a line. A point representing the system where the first particle occupies the position 0, the second position 1, the third position 1, is represented by the triplet 0,1,1.

This representative triplet is **different** from the triplet representing the system where the first particle is in position 1, the second 0 and the third 1, which is given by the point 1,0,1.

In our example we could think of the system of three particles on the line as represented by a point in three-dimensional space, namely by the point whose three coordinates are determined by the positions of the three particles. But this would be a very strange three-dimensional space in which points like (0,1,1) and (1,0,1) had to be regarded as identical! If we wish to represent systems of more than three particles, and not necessarily confined to a single straight line, the situation becomes much more complicated, though the principle is the same.

In quantum theoretical statistical mechanics there are different procedures for counting configurations involving distinguishable and non-distinguishable particles, the statistics of Bose-Einstein and Fermi-Dirac particles. It is as if the space of possible positions was different for one or the other entities.

Space at large, the visible universe of stars and galaxies may possess a hierarchical structure and its global geometry may perhaps be that of an infinitely extended discontinuum. In a truly infinite system there is no unique way of defining homogeneity or approximate homogeneity in time and space as championed by the "steady state universe" proponents. Nor is it obvious how to extrapolate the meaning of the dimensions of physical quantities, including that of the passage of time, to conditions around a singularity which an explosive origin of the universe, "the big bang" picture, involves.

Mathematicians can find a fertile field of study defining various meanings of dimensional analysis and constructing various models for what we call space and time. How much of our picture of the universe is objective and how much conditioned by the structure of our brain and nervous system which are both part of it while analyzing it, remains an intriguing question in the philosophy of science.

CHAPTER 4

PHILOSOPHICAL IMPLICATIONS OF SOME RECENT SCIENTIFIC DISCOVERIES

I WOULD LIKE TO DISCUSS now some discoveries of recent years in mathematics and physics and biology of relevance to my general theme. Some of them are not at all well known, even to philosophers of science. I think these discoveries and inventions have an immense philosophical import, and through that, a moral one as well. I can only try to explain some of them in very general terms in the hope that one can appreciate these scientific developments.

To start with pure mathematics, what I am going to relate supports the statement that even in science itself, contrary to the tenets and beliefs that were prevalent in the nineteenth century, it is not possible to have a complete system confined, as it were, to a fixed number of ideas. It was believed that a system of science could be built once and for all which would give us a complete description of the material universe. In mathematics it was found, rather recently, that this is not even possible for the description of the workings of our minds, and so it seems that we have to take account of a process which continues indefinitely and which has an objective existence and development outside of us. This is, perhaps, in the direction of Whitehead's philosophy of the "process." I am not going, of course, to study here the philosophical questions of what existence means. But, to go a little bit beyond the exact sciences, so-called, one could perhaps say that these developments in mathematics and in logic indicate that no finite number of dogmas, if you wish, may be sufficient to describe forever the material universe, not to mention the human mind itself, as embodied by the brain and not to prescribe in a fixed system the ethics and morals of humanity.

Throughout history, including the Nineteenth century, and the beginning of the present one, mathematicians believed that all rational or mathematical human thought can be codified somewhat as follows: A finite number of symbols and statements called axioms are put down. Some rules of operations are given, once and for all, and new statements are evolved according to these rules. It was believed that every statement which is meaningful (the meaning of meaningful is precisely defined) can be decided by a finite procedure as to its truth or falsity. The famous German mathematician, Hilbert, explicitly propounded this credo. He said mathematics will triumph in the sense that there will be no *ignorabimus*. What actually happened is that around 1930, and even much more recently, mathematical theorems were proved showing the contrary. The mathematician Gödel has shown that every mathematical system which is sufficiently large will contain statements which are sensible within this system but undecidable within it. That is to say such statements will be unprovable by means allowed in the system, but also it will not be possible to disprove them. This is, I think, not only mathematically, but more generally philosophically, of the greatest importance. These statements which Gödel has constructed for any system might have looked arbitrary or perhaps unnatural to some mathematicians. Some might have criticized the results by saying, "Oh well, such statements are meaningful perhaps formally, but nobody would want to know the answers as to their truth anyhow." However, during the last few decades several mathematicians have shown, Paul Cohen especially, that some of the famous old problems of mathematics belong to this class and in the presently used systems of mathematics they are neither true nor false. It seems therefore, that mathematics may be considered perhaps as some other developing process in the living world or in the organic world. It is in the process of progressing and changing in a way that cannot be formulated or predicted in advance, once and for all, ahead of time.

In theoretical physics the theories have also taken a rather unexpected turn. I cannot explain of course what is involved in any coherent and complete way. It was once believed that physical quantities, like the position and the speed of a particle can be measured theoretically at least, with arbitrary precision. According to quantum theory this is not so. One cannot do it simultaneously; there will be unavoidably an error or uncertainty in either one or the other. Of course, for macroscopic phenomena this is so fantastically small that it does not matter. But in the atomic phenomena and in the nuclear realm this leads to a fundamental change of ideas from the classical concepts. These new principles in theoretical physics, in particular the so-called principle of indeterminacy, have a great general philosophical bearing. Some philosophers may even exaggerate the consequences of this uncertainty principle and draw prematurely or even incorrectly, moral conclusions. I cannot elaborate really the theoretical principles involved. In fact any attempt to do so brings to my mind a telephone call I once had from a colleague. He said, "Stan, I had a discussion with students about the difference between a computer and the human brain. I have to tell them more today. Can you quickly tell me over the telephone what are the main points of difference?" I replied that purely operationally there are at least the following: first of all the human brain has at least many thousand times more elements, called neurons, than any present computer. The computer, on the other hand, operates much faster. One hundred thousand times or perhaps a million times faster in a single operation. But the main point is that present day computers still operate by performing one step at a time in sequence, whereas our brains work on thousands, perhaps tens of thousands of channels in parallel, simultaneously, without our being conscious of it of course. There are many more connections between the elements in the brain. The retina of the eye, for instance, has hundreds of thousands of receptors which react and process and combine the visual impressions simultaneously.

To come back to my theme: Before writing on this topic I reread some of the books written by Poincaré, the French mathematician I have mentioned in a previous chapter who wrote at the beginning of this century. He wrote four books destined for the general public, on scientific popularization and philosophy of science. These are unsurpassed masterpieces of presentation of science and also of style. It is interesting to reread them now on several counts. First of all, one is amazed at how much simpler science was just 50 years ago. Daring generalizations were proposed, 50 years ago, much more easily than now, when so many unexpected, strange and weird phenomena are encountered in physics, in astronomy, and in natural sciences in general. Poincaré discussed in one of his books (*Last Thoughts*, published posthumously in 1912) the problem of morality in science. He maintained, and this view was prevalent in those days, that there is no way in which the two, morality and science, can possibly conflict. He said that they are independent and cannot work at cross purposes. Poincaré compared people working professionally in these two domains to soldiers belonging to two different regiments of a single army which has as its aim the battle for the common good of humanity. By the way, he remarked very patriotically, that this resembled the aims of the French army, which has never fought *against* anything, but always *for* something, for ideals. This belief is still held by most scientists: morality and science are never in conflict, and in fact the practice of each has as its aim the progress and the betterment of mankind. But now, the advent of the uses of nuclear energy, the discovery of which was made through science, has instilled doubts in many minds as to such independence. The moral questions arise as to the propriety of scientific work, which may help to produce new weapons. There seems to be a dichotomy or ambivalence in the situation. Leonardo da Vinci did not want to disclose his ideas on the invention of a submarine vehicle for fear that it might be used in war. In a letter to the Duke of Milan, where he offered himself for employment in the Duke's court, he described himself as a painter who could also arrange festivities, make decorations for ceremonies, and could also build fortifications as well or better than anyone else. Was there a difference in his feelings about defensive and offensive weapons? This difference is of course recognized, not only in problems of morality between countries, but in the criminal laws governing individuals in most countries. A homicide in a defensive action against unprovoked attack is considered different from a deliberate homicidal attack. There are of course cases where it may be difficult to distinguish between the two.

There is great diversity in degree of feeling among individuals about the moral propriety of work on weapons and different intensities of feelings of guilt. On one hand one could consider the invention and development of airplanes as tragic since it became the vehicle of most of the tragic misuse of force through weapons. But certainly most do not consider it a diabolic invention.

Again it was Leonardo da Vinci who was among the first to dream about and propose possible constructions of artificial flying birds. Actually, even the most innocuous in appearance, purely theoretical discoveries, can lead ultimately to very evil consequences. Nature seems very strange in caring more about the species than about the individual. This is one of the consequences of Darwin's ideas on evolution and selection. But it is perhaps an essential characteristic of human morality that it should not be completely so. It is hard anyway to describe the purpose or even the meaning of a "win" in the game that nature seems to play between species by natural selection. We can define a win in games like chess or games of cards or games played between two or more groups of individuals, but how to define it in case of competition extending indefinitely into the future between two or more branching chains of individuals? It is clear what a loss means, it is when the chain is wiped out. But the mere survival to infinity for the species does not satisfy our feeling for the meaning of success since then one should say that the plants that we use for food, which presumably will exist as long as the human race, or for that matter the viruses that we cultivate for inoculation, are on par with the coexisting human race. There must be some other as yet undefined criteria for progress in the development. These cannot be found by mere observation of how "nature" has worked so far. The questions of moral purpose will probably be involved in such formulations.

Let us come back to the problem of morality in scientific work. One of the duties of a scientist is to inform those concerned and perhaps the whole public of the possible consequences of their discoveries. Another duty is to teach others and make available all our findings. The immoral thing is to conceal or make false statements. We are all human and we err. But propounding knowingly those statements that are false, and all deliberate falsehoods, all true scientists should and do, not only avoid, but combat. One should certainly allow free expression of all true facts.

Some of the recent developments which I have described, in logic, in mathematics, in physics, show, to state it briefly again, that no fixed set of laws, rules or dogmas will suffice to describe forever the universe as it is, not to mention its future. Activities other than science, like art, which give a different set of unformalized impressions or impulses; or morality and ethics in religion give one, on a different level and in a different way, descriptions of the universe which include the collection of human activities. These broaden the outlook of the scientists and technologists for their very special work. I can only apologize that I cannot make these feelings or ideas clearer and more explicit.

CHAPTER 5

A FIRST LOOK AT COMPUTING:
A PERSONAL RETROSPECTIVE

SOME TWENTY YEARS AGO I made my first visit to the Weizmann Institute, already a world-famous institution. In the short span since its founding, it had made an impressive name for itself, not only in pure science, but also in useful and striking applications to the life sciences and to practical problems of importance.

The occasion for my visit was a meeting at the Institute on problems of hydrodynamics, attended by the late von Karman and other distinguished U.S. physicists. I arrived in Tel Aviv in the evening and need not, I am sure, describe my emotions upon seeing the emblems and symbols of the independent Jewish State. At the airport, to my infinite surprise, I was greeted by a man I had known a quarter of a century earlier in Cambridge, England, Professor Joseph Gillis. Unbelievably, we recognized each other at once, and he drove me to the Institute in what I later learned was one of the very few private cars on campus. In the night air the gardens were redolent with the scent of what I imagined were the plants of the Bible.

Of course, I knew about the work of the Institute's mathematicians and the construction of the "Golem" electronic computer. With great foresight, Israeli scientists such as C.L. Pekeris, Joe Gillis and others, had sensed the significance of this tool for the new State. In Los Alamos, during World War II, John von Neumann had anticipated the potential of electronic computers, the role they could play in the development of science, and their impact on technology. Von Neumann, the man who in the words of physicist Freeman Dyson had "consciously and deliberately set mankind moving along the road that led us into the age of computers," had collaborated in Princeton both with Pekeris and with Gerald Estrin, the two scientists who were to introduce these new ideas to Israel.

The changes brought about by computers have been enormous, not only in science but even more visibly and ubiquitously in industry, especially in the area of communications, and today they pervade daily life in the Western world. It may therefore be of some interest to my reader for me to indicate some of the major advances the use of computers made possible in various scientific fields and also to mention a random sample of the work done on the increasingly more powerful machines that have been and are being developed.

The first meaningful applications were made in problems of mathematical physics, where the complexity of the interaction of many elements may be such that no methods of classical mathematical analysis can give even a qualitative picture of their solution. In hydrodynamics (the study of the motion of fluids) one must have recourse (other than in very simple cases) to numerical work which becomes so massive that only the fastest machines can provide useful approximations. In problems of weather prediction or of the circulation of the atmosphere, initial data also lacks the simplicity and symmetry that allow for the use of theoretical methods of analysis.

Let me point to the so-called Monte Carlo approach as an example of one of the methods that can lead to insight into the behavior of complicated systems. Essentially it consists of imitating physical processes on a computer by following a large but manageable number, replacing the trillions or more that are involved in the real process, by thousands of histories of the behavior of individual particles. This method is used, for instance, in calculations involving nuclear reactors, where neutrons, produced by fission, emerge with various velocities, collide, scatter, and are absorbed or multiply according to probabilities which depend on the initial data. Thus, instead of considering all of the possible eventualities, one selects a number of chains of events which are then followed in a complicated tree of branching eventualities. By surveying the statistics of these "histories" one then gets a good idea of the behavior of the whole system. Without very fast computers, this elaborate procedure would be totally impractical.

Or let us take astrophysics: over the past three decades, scientists have elucidated the evolution of stars during "astronomically long periods of time," i.e. millions of billions of years. Moreover, there have been great advances in our understanding of the dynamics of star clusters and galaxies. Here are two problems which may exemplify one type of possible calculation arising from modern astronomy:

1. Of the visible stars, a high proportion are multiple — double, triple and more — held together by gravitational forces. The process of stellar genesis could perhaps explain this great frequency. Assuming a mass of gas of large dimensions, say on the scale of our entire planetary system, originating as a fluctuation in a perhaps larger mass, one could speculate as to how that mass would contract through gravity. It would be interesting to perform a great number of computations starting with different plausible hypotheses about the initial state of that gas, in order to evaluate the chance that the matter would contract to a single mass in amounts sufficient to produce a star, perhaps with a number of planets, as in the solar system, or that it would contract into a double or multiple system of comparable masses. A vast number of calculations would have to be performed to obtain meaningful statistics for this research.

2. The behavior of a multiple star surrounded by a number of planet-like masses subject to mutual attractions is a classical problem of mechanics. One of the questions in such an n-body problem is: How often will there be physical collisions, that is to say, how often will two of the stars collide, or how often will a "planet" hit one of the stars of a double or triple system? A

number of such calculations could give an approximate idea of the frequency of such collisions and might possibly explain some of the nova stars, those stars which suddenly become exceedingly bright. There are a number of binary stars where novae are observed, sometimes periodically. After all, in our solar system there are periodic collisions of similar chains of meteorites with one member of the system, namely the earth. Of course these are very small bodies, but if there were much larger ones, we would perhaps occasionally observe the considerable brightening of one of the gravitating bodies of the system.

These and similar problems involve so many computations that they can only be performed with the help of very big computers.

In pure mathematics computers have already played a decisive role in certain problems of combinatorics, that is in the study of arrangements and patterns of systems. The famous four-color problem is a shining example. It had long been conjectured that any map drawn on a plane or on the surface of a sphere can be colored using not more than four colors so that countries having a common border will be colored differently. A few years ago, this conjecture was proved on a computer after the general problem had been reduced to one that only involved a finite number of possible standard configurations. The number of these, however, was so great that it took hundreds of hours of computer time to study them.

One could quote many instances of interesting work in pure mathematics, group theory or other areas of algebra in which experimental work has proved useful. The study of non-linear transformations or equations is one field of increasing importance, both in pure mathematics and in applications to physical problems, where heuristic work has already yielded new insights.

Similarly, in the life sciences, the structure of certain organic molecules (among them myoglobin and hemoglobin) was obtained by massive computations to deduce the spatial arrangement of their atoms from the X-ray diffraction spectra. Problems of the interpretation of DNA codes have been reduced, in certain cases, to mathematical analysis. A great deal of work is now in progress to try to understand the meaning of "introns," those longish sequences of symbols in the DNA codes which include the programs for proteins. We now understand thousands of such long strings of symbols, each thousands of units long. The study of their common characteristics and the functions which they initiate pose great challenges.

A list of the advances in science which have only become possible by the use of computers would fill volumes. Let me list just a few examples of the often pioneering work done in the Institute's Applied Mathematics Department during the years of its existence. The electronic energy levels of helium-like atoms have been computed to very high accuracy. Numerical solutions have been obtained of the partial differential equations formulated by Laplace to describe ocean tides, and the solutions agreed remarkably well with observational data. In theoretical seismology, studies have been made of the propagation of seismic waves in a layered spherical earth. In terrestrial spectroscopy, calculations have been made of the periods of free vibrations of the earth on various postulated models of the earth's structure, and comparison with observations could then yield information about the interior of the globe. For the first time accurate numerical solutions of the complete Navier-Stokes equations have been made in fluid dynamics without any approximations or further assumptions. In numerical analysis, efficient methods were developed for the solution of complex differential systems. Biological processes were modeled on the molecular, cellular and population levels.

Also, computer science has stimulated the development of computer logic and the study of problems related to the automatic testing of programs, as well as computer design and construction.

We have indicated how computers have made it possible to determine the structure of proteins and in many cases decipher the meaning of DNA codes. Soon, it may be possible to experiment with modifying, and even perhaps increasing, the speed of evolution.

The study of the higher functions of the human brain, of the influence upon ourselves of our senses, of analogies and associations, of the use of what we call aesthetic criteria, and the ability to reduce complexities to simpler general formulations may all be advanced in the coming decades beyond anything that is now foreseeable.

We should stress in all this the interaction between what our brains can program and experiment with, and what the experiments with these new tools may ultimately do to influence our intuitions and perhaps open up new ways of thinking. It has always seemed curious to me that authors, who write about the future shape of the world, deal with scientific or factual novelties that are only analytic extensions of existing progress. In the realm of science fiction, at least, it is attractive to speculate about changes in the potential of the human brain comparable to the difference which now exists between the brains of other primates and our own. We may also speculate that a vastly increased human memory (with attendant improved mechanisms for access and for the sensing of analogies) is at present undergoing slow qualitative change. The great game involving the history and future trends of the universe — including the organization and evolution of life — will perhaps become a little less mysterious.

CHAPTER 6

COMPUTERS IN MATHEMATICS

ALTHOUGH to many people the electronic computer has come to symbolize the importance of mathematics in the modern world, few professional mathematicians are closely acquainted with the machine. Some, in fact, seem even to fear that individual scientific efforts will be pushed into the background or replaced by less imaginative, purely mechanical habits of research. I believe such fears to be quite groundless. It is preferable to regard the computer as a handy device for manipulating and displaying symbols. Even the most abstract thinkers agree that the simple act of writing down a few symbols on a piece of paper facilitates concentration. In this respect alone — and it is not a trivial one — the new electronic machines enlarge our effective memory and provide a marvelous extension of the means for experimenting with symbols in science. In this chapter I shall try to indicate how the computer can be useful in mathematical research.

The idea of using mechanical or semi-automatic means to perform arithmetical calculations is very old. The origin of the abacus is lost in antiquity, and computers of some kind were evidently built by the ancient Greeks. Blaise Pascal in the 17th century constructed a working mechanism to perform arithmetical operations. Gottfried Wilhelm von Leibniz, one of the creators of mathematical logic as well as the co-inventor of the infinitesimal calculus, outlined a program for what would now be called automatized thinking. The man who clearly visualized a general-purpose computer, complete with a flexible programming scheme and memory units, was Charles Babbage of England. He described a machine he called the analytical engine in 1833 and spent the rest of his life and much of his fortune trying to build it.

Among the leading contributors to modern computer technology were an electrical engineer, J. Presper Eckert, Jr., a physicist, John W. Mauchly, and one of the leading mathematicians of this century, John von Neumann. In 1944 Eckert and Mauchly were deep in the development of a machine known as ENIAC, which stands for Electronic Numerical Integrator and Computer. Designed to compute artillery firing tables for the Army Ordnance Department, ENIAC was finally completed late in 1945. It was wired to perform a specific sequence of calculations; if a different sequence was needed, it had to be extensively rewired. On hearing of the ENIAC project during a visit to the Aberdeen Proving Ground in the summer of 1944, von Neumann became fascinated by the idea and began developing the logical design of a computer capable of using a flexible stored program: a program that could be changed at will without revising the computer's circuits.

A major stimulus for von Neumann's enthusiasm was the task he faced as consultant to the theoretical group at Los Alamos, which was charged with solving computational problems connected with the atomic bomb project. After a discussion in which we reviewed one of these problems von Neumann turned to me and said: "Probably in its solution we shall have to perform more elementary arithmetical steps than the total of all the computations performed by the human race heretofore." I reminded him that there were millions of schoolchildren in the world and that the total number of additions, multiplications and divisions they were obliged to perform every day over a period of a few years would certainly exceed that needed in our problem. Unfortunately we could not harness this great reservoir of talent for our purposes, nor could we in 1944 command the services of an electronic computer. The atomic bomb calculations had to be simplified to the point where they could be solved with paper and pencil and the help of old-fashioned desk calculators.

Down the hall from my office at the Los Alamos Scientific Laboratory was an electronic computer known as MANIAC II (Mathematical Analyzer, Numerical Integrator and Computer), an advanced version of MANIAC I, which von Neumann and his associates completed at the Institute for Advanced Study in 1952. MANIAC II, which was put in operation in 1957, can add two numbers consisting of 13 decimal digits (43 binary digits) in about six microseconds (six millionths of a second). In a separate building nearby is another computer called STRETCH, built by the International Business Machines Corporation, which can manipulate numbers containing 48 binary digits with about 10 times the overall speed of MANIAC II.

MANIAC II and STRETCH are examples of dozens of custom-designed computers built throughout the world in the past 40 years. The first of the big commercially built computers, UNIVAC I, was delivered to the Bureau of the Census in 1951; three years later the General Electric Company became the first industrial user of UNIVAC I. In the years since the first Univac many thousands of computer systems of various makes and sizes have been put to work by the U.S. Government, industry and universities.

Together with increases in arithmetical speed have come increases in memory capacity and in speed of access to stored numbers and instructions. In the biggest electronic machines the memory capacity is now many billions of individual binary digits. I am referring here to the "fast" memory, to which the access time can be much shorter than a microsecond. The time is steadily being reduced; a hundredfold increase in speed seems possible in the near future. A "slow" memory, used as an adjunct to the fast one, normally consists of digits stored on magnetic tape or disk and can be of almost unlimited capacity. The size of memory devices and basic electronic circuits has been steadily reduced, until now even the most elaborate computer can fit into a small room.

It is apparent that many problems are so difficult that they would tax the capacity of any machine one can imagine being built in the next decade. For example, the hydrodynamics of compressible fluids can be studied reasonably well on existing machines if the investigation is limited to problems in two dimensions, but it cannot be studied very satisfactorily in three dimensions. In a two-dimensional study one can imagine that the fluid is confined in a "box" that has been divided into, say, 10,000 cells; the cells are expressed in terms of two coordinates, each of which is divided into 100 parts. In each cell are stored several values, such as those for density and velocity, and a new set of values must be computed for each successive chosen unit of time. It is obvious that if this same problem is simply extended to include a third dimension, storage must be provided for a million cells. One of the studies that is limited in this way is the effort to forecast the weather, for which it would be desirable to use a many-celled three-dimensional model of the atmosphere.

Sometimes when a problem is too complex to be solved in full detail by computer, it is possible to obtain a representative collection of specific solutions by the "Monte Carlo" method. Many years ago I happened to consider ways of calculating what fraction of all games of solitaire could be completed satisfactorily to the last card. When I could not devise a general solution, it occurred to me that the problem could be examined heuristically, that is, in such a way that the examination would at least give an idea of the solution. This would involve actually playing out a number of games, say 100 or 200, and simply recording the results. It was an ideal task for a computer and was at the origin of the Monte Carlo method.

This method is commonly applied to problems of mathematical physics such as those presented by the design of nuclear reactors. In a reactor neutrons are released; they collide, scatter, multiply and are absorbed or escape with various probabilities, depending on the geometry and the composition of the fuel elements and other components. In a complicated geometry no way is known to compute directly the number of neutrons in any given range of energy, direction and velocity. Instead one resorts to a sampling procedure in which the computer traces out a large number of possible histories of individual particles. The computer does not consider all the possible things that might happen to the particle, which would form a very complicated tree of branching eventualities, but selects at each branching point just one of the eventualities with a suitable probability (which is known to the physicist) and examines a large class of such possible chains of events. By gathering statistics on many such chains one can get an idea of the behavior of the system. The class of chains may have to be quite large but it is small compared with the much larger class of all possible branchings. Such sampling procedures, which would be impracticable without the computer, have been applied to many diverse problems.

The variety of work in mathematical physics that has been made possible in recent years through the use of computers is impressive indeed. Astronomy journals, for instance, contain an increasing number of computer results bearing on such matters as the history of stars, the motions of stars in clusters, the complex behavior of stellar atmospheres and the testing of cosmological theories. It has long been recognized that it is mathematically difficult to obtain particular solutions to problems involving the general theory of relativity so that the predictions of alternative formulations can be tested by observation or experiment. The computer is now making it possible to obtain such predictions in many cases. A similar situation exists in nuclear physics with regard to alternative field theories.

I should now like to discuss some particular examples of how the computer can perform work that is both interesting and useful to a mathematician. The first examples are problems in number theory. This subject deals with properties of ordinary integers and particularly with those properties that concern the two most fundamental operations on them: addition and multiplication.

As in so much of "pure" mathematics the objective is to discover and then prove a theorem containing some general truth about numbers. It is often easy to see a relation that holds true in special cases; the task is to show that it holds true in general.

Karl Friedrich Gauss, called "the prince of mathematicians" by his contemporaries, greatly favored experiments on special cases and diligent work with examples to obtain his inspirations for finding general truths in number theory. Asked how he divined some of the remarkable regularities of numbers, he replied, *"Durch planmässiges tattonieren"* — through systematic trying. Srinivasa Ramanujan, the phenomenal Indian number theorist, was equally addicted to experimentation with examples. One can imagine that in the hands of such men the computer would have stimulated many more discoveries in number theory.

A fascinating area of number theory is that dealing with primes, the class of integers that are divisible only by themselves and by one. The Greeks proved that the number of primes is infinite, but even after centuries of work some of the most elementary questions about primes remain unanswered.

For example, can every even number be represented as the sum of two primes? This is the famous Goldbach conjecture. Thus $100 = 93 + 7$ and $200 = 103 + 97$. It has been shown that all even numbers smaller than 2,000,000 can be represented as the sum of two primes, but there is no proof that this holds true for *all* even integers.

It is an interesting fact that there are many pairs of primes differing by two, for instance 11 and 13, 17 and 19, 311 and 313. Although it might seem simple to show that there are infinitely many such pairs of "twin primes," no one has been able to do it. These two unsolved problems demonstrate that the inquiring human mind can almost immediately find mathematical statements of great simplicity whose truth or falsehood are inordinately difficult to decide. Such statements present a continual challenge to mathematicians.

The existence of a proof does not always appease the mathematician. Although it is easily proved that there is an infinite number of primes, one would like to have a formula for writing down an arbitrarily large prime. No such formula has been found. No mathematician can now write on demand a prime with, say, 10 million digits, although one surely exists.

One of the largest known primes was found not long ago with

the help of an electronic computer. Of the form $2^n - 1$, it is called a Mersenne number. There may be an infinite number of primes of this form. No one knows.

Other special numbers that may or may not yield many primes are Fermat numbers, which have the form $2^{2^n} + 1$. For n's of 0, 1, 2 and 3 the corresponding Fermat numbers are 3, 5, 17 and 257. Even for moderate values of n Fermat numbers become extremely large. It is not known, for instance, if the Fermat number with an n of 13 is a prime (the number is $2^{2^{13}} + 1$, or $2^{8192} + 1$).

It is convenient for computer experimentation that both Mersenne and Fermat numbers have a particularly simple appearance when they are written in binary notation. Fermat numbers start with a 1, are followed by 0's and end with a 1. Mersenne numbers in binary notation consist exclusively of 1's. With computers it is an easy matter to study empirically the appearance of primes written in binary form.

The following statement is most likely true: There exists a number n such that an infinite number of primes can be written in a binary sequence that contains exactly n 1's. (The number of 0's interspersed among the 1's, of course, would be unlimited.) Although this statement cannot be proved with the present means of number theory, I suspect that experimental work with a computer might provide some insight into the behavior of binary sequences containing various numbers of 1's. The following experience may help to explain this feeling.

Many years ago my colleague Mark B. Wells and I planned a computer progam to study some combinatorial properties of the distribution of 0's and 1's in prime numbers when expressed in binary form. One day Wells remarked: "Of course, one cannot expect the primes to have, asymptotically, the same number of ones and zeros in their development, since the numbers divisible by three have an even number of ones." This statement was based on the following argument: One would expect a priori that in a large sample of integers expressed in binary form the number of 1's and 0's ought to be randomly distributed and that this should also be the case for a large sample of primes. On the other hand, if it were true that all numbers divisible by three contain an even number of 1's, then the distribution of 1's and 0's in a large sample of primes should not be random.

Returning to my office, I tried to prove Wells's statement about numbers divisible by three but was unsuccessful. After a while I noticed that the statement is not even true. The first number to disprove it is 21, which has three 1's in its binary representation.

Nevertheless, a great majority of the integers divisible by three seem to have an even number of 1's. Beginning with this observation, Wells managed to prove a general theorem: Among all the integers divisible by three from 1 to 2^n, those that have an even number of 1's always predominate, and the difference between their number and the number of those with an odd number of 1's can be computed exactly: it is $3^{(n-1)/2}$. Wells developed corresponding proofs for statements on integers divisible by five, seven and certain other numbers, although he found these theorems increasingly harder to prove.

MERSENNE NUMBER $(2^n - 1)$			FERMAT NUMBER $(2^{2^n} + 1)$		
n	DECIMAL	BINARY	n	DECIMAL	BINARY
1	1	1	0	3	11
2	3	11	1	5	101
3	7	111	2	17	1001
4	15	1111	3	257	100000001
5	31	11111	4	65,537	10000000000000001

MERSENNE AND FERMAT NUMBERS have a simple appearance when written in binary notation. Although many Mersenne numbers are not prime (for example 15), there may be an infinite number of primes of this form. There may also be an infinite number of Fermat primes, but even the Fermat number for an *n* as small as 13 has not yet been tested.

3	11	27	11011
6	110	30	11110
9	1001	33	100001
12	1100	36	100100
15	1111	39	100111
18	10010	42	101010
21	10101	45	101101
24	11000	48	110000

INTEGERS DIVISIBLE BY THREE usually contain an even number of 1's when written in binary form. This observation led to the proof of a general theorem, described in the text.

By now quite a few problems in number theory have been studied experimentally on computers. Not all of this work is restricted to tables, special examples and sundry curiosities. D.H. Lehmer of the University of California at Berkeley has made unusually effective use of the computer in number theory. With its help he has obtained several general theorems. Essentially what he has done is to reduce general statements to the examination of a large number of special cases. The number of cases was so large that it would have been impracticable, if not impossible, to go through them by hand computation. With the help of the computer, however, Lehmer and his associates were able to determine all exceptions explicitly and thereby discover the theorem that was valid for all other cases. Unfortunately Lehmer's interesting work is at a difficult mathematical level and to describe it would take us far afield.

It must be emphasized that Lehmer's theorems were not proved entirely by machine. The machine was instrumental in enabling him to obtain the proof. This is quite different from having a program that can guide a computer to produce a complete formal proof of a mathematical statement. Such a

1,2	1,3	1,4	1,5	1,6	1,7
	2,3	2,4	2,5	2,6	2,7
		3,4	3,5	3,6	3,7
			4,5	4,6	4,7
				5,6	5,7
					6,7

1,2,3	1,3	1,4,5	1,5	1,6,7	1,7
	2,3	2,4,6	2,5,7	2,6	2,7
		3,4,7	3,5,6	3,6	3,7
			4,5	4,6	4,7
				5,6	5,7
					6,7

STEINER PROBLEM poses this question: Gives n objects, can they be arranged in a set of triplets so that every pair of objects appears once and once only in every triplet? The problem can be solved only when $n = 6k + 1$ or $6k + 3$, in which k can be an integer. One solution for $k = 1$, in which case $n = 7$, is shown here. The first table lists all possible pairs of seven objects. The second table shows seven triplets that contain all pairs only once. The 21 digits in these triplets can be regrouped into other triplets.

program, however, is not beyond the realm of possibility. The computer can operate not only with numbers but also with the symbols needed to perform logical operations. Thus it can execute simple orders corresponding to the basic "Boolean" operations. There are essentially the Aristotelian expressions of "and," "or" and "not." Under a set of instructions the computer can follow such orders in a prescribed sequence and explore a labyrinth of possibilities, choosing among the possible alternatives the ones that satisfy, at any moment, the result of previous computations.

With such techniques it has been possible to program a computer to find proofs of elementary theorems in Euclid's geometry. Some of these efforts, particularly those pursued at the International Business Machines Research Center, have been quite successful. Other programs have enabled the computer to find proofs of simple facts of projective geometry. I have no doubt that these efforts mark only the beginnings; the future role of computers in dealing with the "effective" parts of mathematics will be much larger.

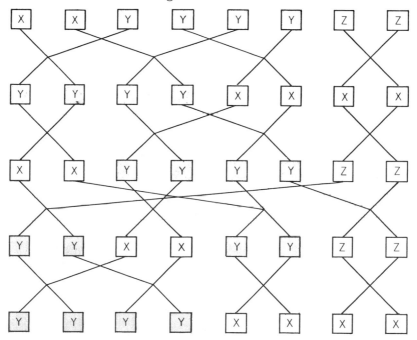

GENEALOGICAL "TREES" raise many interesting combinatorial questions. In the simple case shown here individuals of three different colors mate in pairs. Strictly, x, y and z specify the fraction of each color in each generation, but here they also identify color type. Each mating produces a pair of offspring and the color of the offspring is uniquely determined by the colors of the parents according to a fixed rule. (For example, 2 y's or 2 z's produce 2 x's.) Assuming an initial population containing hundreds of members, one might ask such questions as these: Given an individual in the fifth generation, how many different ancestors does he have in, say, the first generation? What are the proportions of x's, y's and z's among the ancestors of a given individual in the nth generation?

I shall now turn from the study of integers to combinatorial analysis and discuss some of the uses of the computer in this field. Very briefly, combinatorial analysis deals with the properties of arrangements and patterns defined by means of a finite class of "points." Familiar examples are the problems on permutations and combinations studied in high school algebra. In a typical case one starts with a finite set of n points and assumes certain given, or prescribed, relations between any two of them or, more generally, among any k of them. One may then wish to enumerate the number of all possible structures that are related in the prescribed way, or one may want to know the number of equivalent structures. In some cases one may consider the finite set of given objects to be transformations of a set on itself. In the broadest sense one could say that combinatorial analysis deals with relations and patterns, their classification and morphology. In this field too electronic computers have proved to be extremely useful. Here are some examples.

Consider the well known problem of placing eight queens on a chessboard in such a way that no one of them attacks another. For an ordinary 8×8 chessboard there are only 12 fundamentally different solutions. The mathematician would like to know in how many different ways the problem can be solved for n queens on an $n \times n$ board. Such enumeration problems are in general difficult but computer studies can assist in their solution.

The following problem was first proposed in the 19th century by the Swiss mathematician Jakob Steiner: Given n objects, can one arrange them in a set of triplets in such a way that every pair of objects appears once and only once in a triplet? If n is five, for example, there are 10 possible pairs of five objects, but a little experimentation will show that there is no way to put them all in triplets without repeating some of the pairs. The problem can be solved only when $n = 6k + 1$ or $6k + 3$, in which k is any integer. The solution for $k = 1$ (in which case $n = 7$) is shown at the top of page 52. The number of triplets in the solution is seven. In how many ways can the problem be solved? Again, the computer is very useful when k is a large number.

The shortest-route problem, often called the traveling sales-man problem, is another familiar one in combinatorics. Given are the positions of n points, either in a plane or in space. The problem is to connect all the points so that the total route between them is as short as possible. Another version of this problem is to find the route through a network of points (without neces-sarily touching all the points) that would take the minimum time to traverse.

These problems differ from the two preceding ones in that they necessitate finding a method, or recipe, for constructing the minimum route. Strictly speaking, therefore, they are problems in "meta-combinatorics." This term signifies that a precise formulation of the problem requires a definition of what one means by a recipe for construction. Such a definition is possible, and precise formulations can be made. When the n points are distributed in a multi-dimensional space, the prob-lem can hardly be tackled without a computer.

A final example of combinatorics can be expressed as a problem in genealogy. Assume, for the sake of simplicity, that a population consists of many individuals who combine at random, and that each pair produces, after a certain time, another pair. Let the process continue through many genera-tions and assume that the production of offspring takes place at the same time for all parents in each generation. Many in-teresting questions of combinatorial character arise immediately.

For instance, given an individual in the 15th generation of this process, how many different ancestors does he have in, say, the ninth generation? Since this is six generations back it is obvious that the maximum number of different ancestors is 2^6, but this assumes no kinship between any of the ancestors. As in human genealogy there is a certain probability that kinship exists and that the actual number is smaller than 2^6. What is the probability of finding various smaller numbers?

Suppose the original population consists of two classes (that is, each individual has one or the other of two characteristics); how are these classes mixed in the course of many generations? In other words, considering any individual in the nth generation, one would like to know the proportion of the two characteristics among all his ancestors.

Let us now make a slightly more realistic assumption. Consider the process as before but with the restriction removed that all offspring appear at the same time from parents of the same age. Assume instead that the production of the new generation is spread over a finite period of time according to a specific probability distribution. After this process has continued for some time the individuals of the most recent generation will be, so to speak, of different generations. A process of this kind actually occurs in human populations because mothers tend to be younger, on the average, than fathers. Therefore going back, say, 10 generations through the chain of mothers yields a smaller number of total years than going back through the chain of 10 fathers. It becomes a complex combinatorial problem to calculate the average number of generations represented in the genealogical history of each individual after many years have elapsed from time zero. This and many similar questions are difficult to treat analytically. By imitating the process on a computer, however, it is easy to obtain data that throw some light on the matter.

The last mathematical area I should like to discuss in connection with computers is the rather broad but little-explored one of non-linearity. A linear function of one variable has the form $x' = ax + b$, where a and b are constants. Functions and transformations of this form are the simplest ones mathematically, and they occur extensively in the natural sciences and in technology. For example, quantum theory employs linear mathematics, although there are now indications that future understanding of nuclear and subnuclear phenomena will require non-linear theories. In many physical theories, such as hydrodynamics, the equations are non-linear from the outset.

The simplest non-linear functions are quadratic; for one variable such functions have the form $y = ax^2 + bx + c$, where a, b and c are constants. It may surprise non-mathematical readers how little is known about the properties of such non-linear functions and transformations. Some of the simplest questions concerning their properties remain unanswered.

As an example, mathematicians would like to learn more about the behavior of non-linear functions when subjected to the process known as iteration. This simply means repeated application of the function (or transformation) to some starting value. For instance, if the point described by a function is the square root of x, the iteration would be the square root of the square root of x; each succeeding iteration would consist of again taking the square root.

A transformation given by two functions containing two variables each defines a point on a plane; its iteration gives rise to successive points, or "images." Finding the properties of the sequence of iterated images of a single point, when described by a non-linear function, is in general difficult. Present techniques of analysis are inadequate to unravel the behavior of these quite simply defined transformations fully.

Here again empirical work with the computers can be of great help, particularly if the computer is equipped to display visually the location of many iterated points on the face of an oscilloscope. This enables us to see at a glance the results of hundreds of iterations.

In examining such displays the mathematician is curious to learn whether or not the succession of iterated images converge to a single location, or "fixed point." Frequently the images do not converge but jump around in what appears to be a haphazard fashion — when they are viewed one by one. But if hundreds of images are examined, it may be seen that they converge to curves that are often most unexpected and peculiar. Such empirical work has led my associates and me to some general conjectures and to the finding of some new properties of non-linear transformations.

What are the obvious desiderata that would make the electronic computer an even more valuable tool than it is today? One important need is the ability to handle a broader range of logical operations. As I have noted, the simplest operations of logic, the Boolean operations, have been incorporated in electronic computers from the outset. In order to encompass

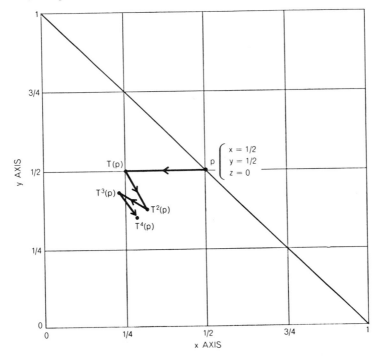

$$x' = y^2 + z^2 \qquad y' = 2xy + 2xz \qquad z = x^2 + 2yz$$

INITIAL POINT p	FIRST ITERATION T(p)	SECOND ITERATION T²(p)	THIRD ITERATION T³(p)	FOURTH ITERATION T⁴(p)
x = 1/2	x' = 1/4	x" = 5/16	x''' = 61/256 = .238	x'''' = .295
y = 1/2	y' = 2/4	y" = 6/16	y''' = 110/256 = .430	y'''' = .363
z = 0	z' = 1/4	z" = 5/16	z''' = 85/256 = .332	z'''' = .342
SUM 1	1	1	1 = 1.000	1.000

PROCESS OF ITERATION involves repeated applications of a function (or transformation) to an initial value or a point. Here three equations containing three variables define a point in a plane. Iteration gives rise to successive points, or "images." Because the three variables always add up to 1, only two variables (say x and y) need to be plotted. The first iteration, $T(p)$, is obtained by inserting initial values of x, y and z ($\frac{1}{2}$, $\frac{1}{2}$, 0) in the three equations. The new values x', y', z' ($\frac{1}{4}$, $\frac{1}{2}$, $\frac{1}{4}$) are then inserted to produce the second iteration, $T^2(p)$, and so on. Computers can quickly compute and display thousands of iterations of a point so that their behavior can be studied.

more of contemporary mathematics the computer needs a "universal quantifier" and an "existential quantifier." The universal quantifier is required to express the statement one sees so frequently in mathematical papers: "*For all x* such and such holds." The existential quantifier is needed to express another common statement: "*There exists* an *x* so that such and such is true." If one could add these two quantifiers to the Boolean operations, one could formulate for computer examination most of traditional and much of modern mathematics. Unfortunately there is no good computer program that will manipulate the concepts "for all" and "there exists."

One can take for granted that there will be continued increases in processing speed and in memory capacity. There will be more fundamental developments too. Most present computers operate in a linear sequence: they do one thing at a time. It is a challenge to design a machine more on the model of the animal nervous system, which can carry out many operations simultaneously.

One can imagine new methods of calculations specifically adapted to the computer. Thanks to the speed of the machine one will be able to explore almost palpably, so to say, geometrical configurations in spaces of more than three dimensions and one will be able to obtain, through practice, new intuitions. These will stimulate the mathematician working in topology and in the combinatorics of new mathematical objects. These objects may be ordinary integers, but integers far exceeding in size and number any now used for experimentation. One should also be able to develop mathematical expressions with many more existential quantifiers than are now employed in formal mathematical definitions. New games will be played on future machines; new objects and their motions will be considered in spaces now hard to visualize with our present experience, which is essentially limited to three dimensions.

The old philosophical question remains: Is mathematics largely a free creation of the human brain, or has the choice of definitions, axioms and problems been suggested largely by the external physical world? (I would include as part of the physical world the anatomy of the brain itself.) It is likely that work with electronic machines, in the course of the next decade or so, will shed some light on this question. Further insight may come from the study of similarities between the workings of the human nervous system and the organization of computers. There will be novel applications of mathematics in the biological sciences, and new problems in mathematics will be suggested by the study of living matter.

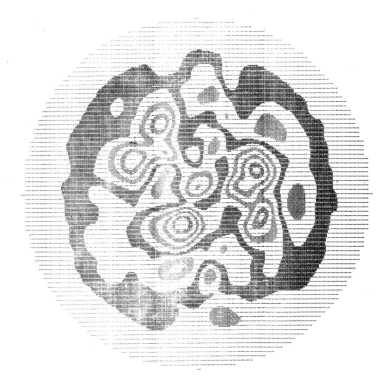

SIMULATED WEATHER PATTERN for the entire Northern Hemisphere was produced by the **STRETCH** computer at the General Circulation Research Laboratory of the U.S. Weather Bureau. In an effort to develop and test new theories of atmospheric behavior, investigators program the computer with equations that attempt to account for atmospheric phenomena. Data from real observations are then fed in, from which the computer produces changing model weather patterns for days and weeks. These are compared with actual observations over the period. The shaded "contours" on this map show simulated sea-level atmospheric pressure during one of these studies. The pattern is built up entirely of densely packed numbers and letters printed out directly by the computer itself.

Chapter 7

EXPERIMENTS IN CHESS
ON ELECTRONIC COMPUTING
MACHINES: SOME EARLY EFFORTS

with P. R. Stein

Is CHESS AN ART or a science? All devotees of the game know that the great player is one who possesses superior "insight" — but what, precisely, is its nature? One view is that chess skill is a divine gift; some players seem to possess an intuitive feeling for the pregnant situation, amounting at times to an almost mystical perception. At the other extreme lies the theory that great chess is the result of consistent application of rational principles. Most writers on the game will be found to have taken their stand somewhere in between. They speak, indeed, of "rational principles," but when it comes to the question of "consistent application" they throw up their hands and, at least by implication, support the intuitionists. One suspects that many authorities regard the "rational principles" of chess as somehow inferior, indispensable but prosaic guides to lead the player through the maze of complications when his intuition is unhappily dormant. Alternatively, one can very well argue in reverse, that chess is basically an exercise in logic and that intuition is only called upon when the process of calculation becomes too difficult or, at least, too time-consuming.*

Our knowledge of chess to date tends to support the view of

---

* It is hardly necessary to mention that, mathematically speaking, chess is in principle completely determined, that is, a well-defined strategy exists which leads to a unique outcome. Practically, of course, the determination of this strategy is not feasible.

the game as more an art than a science. Only in certain standard situations e.g., the simplest end-games and mates, can chess be thought of as reduced (some die-hards might prefer the word "elevated") to a science in the sense that a certain well-defined procedure will invariably produce the expected result. Despite many decades of intensive study and analysis, the "correct" handling of the opening remains an art rather than a science. Innovations are constantly appearing as early as the fourth or fifth move in what were previously thought to be the best established lines.

"Best Play" Theories

As for the middle game, technique in this phase of play can hardly be described logically, except in the most general terms, and illustrated, paradoxically, by elucidation of a few very special situations. Yet no one can deny that there must exist, in theory, principles capable of more or less precise formulation of what constitutes the basis of "best play." The first, and in some ways the most thoroughgoing attempt to enunciate such principles is that of Wilhelm Steinitz.

Steinitz's theory is discussed at length in Dr. Emanuel Lasker's famous book "Manual of Chess." Lasker was a mathematician by profession and one would not be surprised to see him exalt the rational element in chess. Curiously enough, while praising reason in general, and Steinitz's theories in particular, Lasker seems to have felt that the net result was insufficient. That is, while a knowledge of the rational principles laid down by Steinitz is indispensable, it is not enough. One needs imagination and, above all, the ability to act in the face of the unforeseen, both in the chess "struggle" and in life, of which he considered chess to be a sort of purified miniature. (See Reinfeld's excellent introduction to Lasker's book.)

One feels that Dr. Lasker would not have applauded efforts to reduce chess to a set of essentially numerical rules. Steinitz's data concerning the attack and the defense, and above all, the principle of the "accumulation of small permanent advantages," have been the basis of all later rational developments. On the practical side, certain concepts such as rapid development, weak squares, open files, etc., have shown themselves to be highly relevant in the majority of situations.

These concepts are certainly basic to the analysis of positions as well as being indispensable guides in actual play. They, at least, have a statistical validity. (We might add that they are the meat and drink of the commentators; without them, the art of annotation would consist solely in the listing of continuations.)

These precepts give rise to criteria by which the relative strengths of certain types of positions may be judged. Do they, then, constitute a set of "principles" for the play of the game? It is permissible to doubt it. There must be some more fundamental ideas involved. The authors are not prepared to say what these are, but at least two obvious postulates must be included. The first and most obvious is the principle of "Material Advantage" i.e., the occurring and maintaining of a plus in material. This point needs no discussion; specific tactical considerations apart (and certainly *"ceteris paribus,"* as Lasker would say), it is clearly a good thing to be a piece up.

Secondly, there is the principle of "Superior Mobility," i.e., the securing of more freedom for one's pieces and the restriction of the opponent's field of action. This principle can be seen in operation even in the play of the average chess amateur. No one can doubt the validity of these elementary desiderata: the interesting question is how far method of play can be based on these two principles alone. What then about combination? What about direct mating attacks?

We think that such tactical points are to some extent taken care of in the application of the two general principles. (A combination can be looked on, roughly, as a device to win material or to gain room for maneuver.) As for mating attacks, the mate is the most extreme case of immobility for the player mated (the case of stalemate, of course, requires special consideration).

Programming the Code

These simple thoughts appeared to merit some sort of test. Accordingly, a group of scientists at Los Alamos, including amateur chess players, decided in the mid-fifties to construct a program which would enable an electronic computing machine to play chess utilizing these two criteria of material advantage and mobility.

Let it be said once and for all that computing machines do not "think." They add, subtract, multiply and divide, in addition to which they can make elementary decisions: e.g., whether or not a given number is larger, smaller, or equal to a second given number. All more complicated operations must be compounded from these simple ones.

To make a computing machine perform any given task, all the rules must be provided in advance, and these in quite explicit form. The more complicated the problem, the more detailed and elaborate must be the code embodying the rules of calculation. In a game like chess the computer must proceed in a quite naïve fashion, essentially trying all possibilities which are allowed by the rules of the game, and then picking the "best" in accordance with some pre-arranged criterion. Starting from a given chess position, a sequence of moves of prescribed length can very well include millions of possibilities.

64 Million Sequences

For instance, if we consider a sequence four moves long, (two moves by each side), and assume for the sake of argument that there are 20 legal moves possible at each stage, then the number of different possible sequences (in the following we shall call them "chains") is 160,000. If we wish to consider three moves by each player, keeping the number of legal moves available equal to 20, then some 64 million chains can occur!

Lest the reader feel that 20 is too high a figure for the number of possible legal moves at each stage of such a chain, we give a concrete example. Starting from a position reached in a standard line of the King's Indian after seven moves on each side, the average number of legal moves at each stage of the six-move chain (three moves by each side) following the actual column is just over 31. (These are the legal moves, not necessarily good or even sensible moves.) There are thus roughly 890 million chains possible.

In human play, all but a very few of these chains are rejected almost instantaneously and the remaining continuations are subjected to relatively careful scrutiny. How the human brain does this is certainly a great mystery; and, until we know more about the mechanism of thought, we cannot incorporate such features into a machine code. That the machine can achieve any worthwhile result by its naïve method of trial and error is due to the great speed with which it performs its elementary operations.

Thus we see that even with their great advantage of speed the length of the chain that can be considered by a computer at any stage of a chess game is severely limited by time considerations. In addition, if the machine had to "remember" too much information during the course of the game, it might well run out of space in which to store this information; another way of putting this is to say that its "memory" capacity might be exceeded. In practice, this is usually not so serious a limitation as that imposed by the time required per move, since the amount of information that must be retained from move to move can be reduced by ingenious coding.

On the other hand, if one should wish to incorporate a large set of rules applying to special situations that arise in practice, one might well overtax the memory of even the largest computer. For example, suppose that our main purpose were to make a computing machine play a better-than-average game of chess. In view of the impracticability of extending the "range of foresight" it might seem a good idea to incorporate into the machine's memory several thousand typical positions (arranged perhaps in some thematic order.)

The machine could then, by direct comparison, choose that pattern most relevant to the actual position, after which it could consult a library of suggested continuations and tentatively select its move, subjecting the result to some standard evaluation. The machine would then be playing, so to speak, by analogy. To do so properly might require a tremendous amount of storage space. Actually we do not feel that such an approach is fruitful as yet, less because of the technical difficulties involved than because the results would not shed much light on the fundamental structure of the game.

Try "6 × 6" Chess

With these considerations in mind, we decided to test our ideas, not on the classical game of chess, but rather on a somewhat simplified version which we call "6 × 6 chess." This is a game in which the Bishops and Bishop Pawns are omitted and the rest of the pieces are arranged in the usual order on a 6 × 6 board. The rules of the game are identical with those of ordinary chess except that castling is not allowed, and the initial Pawn advance is restricted to one square. (The reason for this last rule becomes obvious on setting up the board.)

It turns out that 6 × 6 chess is an interesting game in its own right, though vastly simpler than its 8 × 8 parent. Much of the flavor of chess is retained, though it soon becomes clear that there is little room for maneuver. The reason we chose this game for our initial experiments was to enable a machine to look two moves ahead, i.e., two by each side, and still make its moves in a reasonable time (the number of legal moves at each stage is roughly half the number for classical chess). Therefore we get a factor of 16 = 2^4 for chains, and the machine could make a move in about 10 minutes; on the average 8 × 8 game it would take over one hour per move.

We did not use the fastest machine available at that time in our laboratory, but rather a slower and older machine known officially as MANIAC I. This computer could perform about ten thousand elementary operations per second and was ideally suited to the simplified problem we wished to study. Omitting technical details, we need only say that the machine examined all legally allowed four-move chains (i.e. two moves by each player), finally selecting that continuation which resulted in the greatest set mobility, coupled with the greatest material advantage. (This description of the procedure is not strictly accurate, but the actual prescription is too involved to be discussed here.)

Mobility was defined for this as the number of legally available squares; certain weightings were applied later in order to avoid a few obviously senseless moves (cf. White's first move in Game I). Material strength was evaluated by assigning to each Pawn the value of eight legal moves and valuing the other pieces accordingly, following the scale generally accepted as applying to the classical game (the correct relative values are undoubtedly different for our game; as yet they are unknown).

THESE CRITERIA are admittedly extremely crude. Nevertheless, the actual games revealed several amusing features. The first game pitted the machine against itself. The very first move indicated that the code needed revising.

GAME 1

Maniac I.
White

Maniac I.
Black

1 N–K3		P–Q3

White did not "expect" . . . P–Q3. His shortsightedness results in a loss of time. Later, this weakness was corrected.

2 N–N1		P–QN3
3 P–K3		P–K3
4 Q–K2		P–Q4

White continues to waste time.

5 PxP		NPxP
6 Q–Q1		Q–R3

Black gives up his advantage; he is afraid of the check at R3.

7 Q–R3†		QxQ
8 NxQ		K–Q1

Black avoids being checked at N5.

9 N–R3		K–Q2

Black leaves the Pawn on prise.

10 R–QN1		. . . .

White fails to take it. He feels his Queen Knight is not secure.

† = check; ‡ = dbl. check; ! = dis. ch.

10		K–K1
11 N–Q4†		NxN
12 PxN		R–Q1
13 R–N5		. . . .

White wins a Pawn, anyway.

13		N–Q2
14 RxP		R–QN1

Black threatens mate on the move.

15 R–N5		R–R1
16 P–R3		P–R3
17 NxNP		KxN
18 RxN†		K–N1
19 R–K5?		R–K1??

Here Black failed to see the win: 19 . . . RxP 20 P–N3 (forced), P–R4! 21 any, R–R6 mate.

20 RxR†		KxR
21 P–Q5†		. . . .

White's play is hardly precise.

21		KxP
22 K–Q1		R–KN1
23 P–N3		KPxP
24 PxP		R–N3?

Black will not play 24 . . . PxP for fear of the Rook check.

25 RxP		RxR
26 PxR		K–N2

And now it's all over.

27 P–R5		K–R3
28 P–R6(Q)		KxP
29 Q–K4		. . . .

And mate the next move.

BY MINOR ADJUSTMENTS in the evaluation procedure, we were able to correct some of the more serious weaknesses in the machine's play which the first game exposed. We now felt ready to let MANIAC play against a suitably handicapped human player. Dr. Martin Kruskal, of Princeton, graciously accepted the challenge. Since Dr. Kruskal is a strong (8 × 8 chess) player, we felt it only fair to have him give the machine odds of a Queen. As a slight compensation, he was allowed to play White. The game follows.

GAME II

(Remove White's Queen)

Dr. M. Kruskal	Maniac I
White	Black

| 1 P–K3 | N–QR3 |
| 2 P–QN3 | N–N1 |

Black's shortsightedness has not yet been cured.

3 P–Q3	P–K3
4 P–Q4	PxP
5 KPxP	P–QR3
6 N–QR3	P–KN3
7 K–K2	K–N2
8 P–R3	P–N4†
9 K–Q2	PxP†
10 PxP	N–KR3
11 N–K3†	NxN
12 KxN	. . . .

White has succeeded in closing the position so that Black's Queen has no scope. If the Queen ever gets loose, White might as well resign. Black can, of course, free himself by sacrificing his Knight for a Pawn, but such subtlety is well beyond him.

| 12 . . . . | K–R3 |
| 13 K–Q2 | . . . . |

White wants to make room for the Knight at K3. This plan occurred to him just as he was on the verge of conceding a draw. Note that, if he plays 13 N– Q2, then 13 . . . K–N2! forces the Knight back to R3: e.g., 14 RxR? Q–N3†! 15 PxQ (forced), PxP mate.

13	R–R2
14 K–N2	R–N1
15 KR–K1	. . . .

White prevents the dangerous advance, 15 . . . P–K4.

| 15 . . . . | R–KN2 |

This weak move gives White an idea for a mating trap.

16 N–Q2	RxR
17 KxR	R–N1
18 K–N1!	. . . .

This waiting move is the key to the combination. Kruskal thinks he can predict the Machine's next move.

| 18 . . . . | R–N2?? |

White has guessed correctly. Black now loses his Queen.

| 19 N–N1! | Q–N3† |

Black's last move is heart-breaking, but necessary. In eminently human fashion, the machine "deliberated" 20 minutes before giving up its Queen. But, otherwise, 20 N–K3 is mate.

| 20 PxQ | PxP |
| 21 K–Q2 | P–K4† |

Black's inherent limitations afford him no chance in the end-game.

22 PxP	P–N5	30 NxR	K–Q2
23 R–R1	R–N1	31 K–K3	K–N3
24 R–R5	PxP†	32 K–K4	NxP
25 K–K2	R–N3	33 PxN	K–N4
26 RxP	R–N1	34 P–Q5	K–N3
27 NxP†	K–N3	35 P–Q6(Q)	K–N4
28 KxP	K–K3§	36 N–K3	K–R5
29 R–N5	RxR†	37 Q–Q3	K–N6
		38 Q–N2 mate	

THE THIRD GAME matched the machine against a beginner with one week's experience, who had, in fact, been taught the game expressly for the purpose of participating in this contest. The play was naturally weak on both sides, but as the ending shows, MANIAC enjoyed a slight advantage in precision.

GAME III

Maniac I	Beginner
White	Black

1 P–K3	. . . .

An improvement over the opening move of Game I

1	P–QN3	4 N–N1	P–QR3
2 N–KR3	P–K3	5 PxRP	NxP
3 P–QN3	P–N3	6 K–K2?	. . . .
6		N–Q4?	

Analysis indicates that 6 . . . P–Q3 wins. If 7 PxP, QxP, most of the winning lines start with . . . N–KR3 or, when that fails (as on 8 K–K1), with . . .N Q†, etc. If White doesn't take the Pawn, then . . . P–Q4 decides quickly. White's sixth move is certainly very weak; but Black lets him get away with it.

7 NxN	NPxN†

7 . . . KPxN1 is much better.

8 K–K1	P–R3	12 Q–R3	Q–N2
9 P–QR3	R–N1	13 Q–R2†	K–N2
10 P–R4	R–R1	14 R–N1	RxP
11 P–R5	K–K2	15 RxQ	RxQ
		16 R–N1	R–QR2

White's Queen-side demonstration has resulted in the loss of a Pawn; but his general play conveys at least as good an impression as does Black's aimless wandering. Now White starts action in earnest on the King side.

17 P–R3	R–R3?

Black is oblivious to the danger. 17 . . . P–N4 is necessary.

18 RPxP	P–Q3	21 PxR(Q)	N–Q2
19 N–R3†	K–K1	22 QxP†	K–Q1
20 P–N5†	K–K2	23 N–N5 mate	

All things considered, the machine handled the final attack most impressively.

WE MUST NOW ASK, what do these simple experiments indicate for the future attempts? It appears that these games, though weak, tend to confirm the over-ruling importance of our two simple criteria. One must remember that our code had absolutely no "experience," tactical or strategic, built into it. One might think that many of the elementary blunders could be corrected by allowing the computer to explore certain special chains of, for example, exchanges farther than our standard two moves.

One result of this procedure should be to make almost three of four-move mating combinations trivial, even in the 8x8 game. We feel confident, for instance, that the machine will find Marshall's famous "golden move" (Marshall-Levitsky, Breslau, 1912), starting from the same position (and without any coaching from the sidelines!).

In defining mobility we should also not merely count available moves but weight them differently, e.g. moving a Knight closer to the enemy's camp should count more than retracting it, etc.

I feel strongly that, perhaps in contrast to the popular view, it is not the "brilliancies", i.e. on the whole, forced tactical combinations, but their preparation — the general feeling for strategy — which constitutes the deepest part of the game and the most difficult to incorporate in any future machine code.

This factor manifests itself with the greatest clarity in the games of Alekhine. Cf. the comment of Spielmann on Alekhine games: "I see well enough the combinations themselves, but where he gets the positions enabling him to start the combinations is beyond me."

A second look at our 6x6 games, however, reveals a truly shocking weakness. In our comments on Game III, we have mentioned that White's sixth move leads to a lost game. It is quite likely that the exchange on the rook file initiated by White's fifth move is already sufficient to doom him — the unnecessary freeing of Black's King Knight is fatal.

It seems that in 6x6 chess, control of the center plus the slightest freedom to maneuver is decisive, mainly because the adversary is almost necessarily in "Zugzwang" on the crowded board. While it is easy enough to correct the particular fault of White's fifth move by a slight adjustment, it is by no means clear how to avoid the evil via the invocation of a precise and general rule. (Of course, the principle of mobility would be sufficient if coupled with an extended range of foresight.)

Chapter 8

COMPUTATIONS IN PARALLEL

I SHALL TRY to elaborate in a few examples, the desideratum of having machines which perform simultaneously and independently on many channels simple arithmetic and logical operations. I want to discuss briefly the advantages of such a system compared to the now prevailing computers which do their work in series. I shall content myself, in this chapter, with a discussion of the uses of such machines without going into technical schemas of their nature and characteristics.

It seems fairly obvious that one of the reasons for the superior performance of the human brain as compared with the fastest computing machines lies in its ability to gather information and probably to process much of it by working on many channels at once. It is in playing a game like chess that it is most obvious that the visual impression and conception of the board and positions is not obtained by scanning in linear order the information given on a two-dimensional pattern! It is sufficient to compare the quality of the chess games as· played by the computer (cf. Chapter 7) with the human performance to realize the advantages of the latter. The computer examines several hundred thousand cases and evaluates each one, and even with its speed of operation, surpassing that of the brain by many thousands, takes several minutes to look just a few moves ahead. Somehow, even an average chess player is able to reject most

of the chains at a glance, look farther ahead, and select interesting lines. An evaluation of the visual impression involving
many independent points on the retina is performed then on
many channels at once. Probably some of the obvious superiority
of the human brain even in this sort of work is due to its
enormous memory capacity and a very intelligent "super code"
for the qualities of a position. Nevertheless, it is at least the
writer's belief that most of the "intelligence" is due to the facility
for operation in parallel. No doubt the syllogisms and the purely
logical execution of the final product which is made in reasoning,
is performed, to a large extent, in serial order; but the finding
of the chain of action or of the proof of a theorem, if one wants,
involves very likely a number of "independent" explorations
undertaken in parallel at the same time. In other words, the
process of thinking out a mathematical investigation takes place
to a large extent by parallel searches; the final checking and
writing out of a finished product — a theorem with its proof —
is largely linear.

There are examples of mathematical techniques where the
need and advantages of a parallel computation are very obvious.
For the bulk of the calculations in the so-called Monte Carlo
method, one could proceed by performing simultaneously many
random walks and have independent random decisions involving
the fate of each separate particle. When one studies a diffusion
process or a diffusion combined with multiplication and
absorption, e.g., in a problem involving neutron distribution
in a fissile medium, one clearly has to study a number of
independent, or almost independent, histories of individual
particles. An analogous statement can be made about any linear
problem. When the differential or integro-differential equations
describing the behavior of a system are linear in the dependent
variables — be it a system of ordinary differential equations or
a partial differential equation — it is fairly clear that a computation can proceed independently for as many "particles" or "spatial
zones" as were decided upon for the given resolution in the
independent variables.

Now these remarks intend to go beyond the trivial observation that it is better to have many computing machines than just one: the point is that each "particle" or "zone" constitutes a very simple logical situation and the independent parts of the machine would involve only additions and simple switches. If one thinks of a Monte Carlo type problem, each particle would only have to indicate one of a certain number of possible "states" — with an ordinary shift being almost sufficient to perform the changes in state. Or in other words, whatever computations are involved for any single particle, they can be done with very little accuracy, i.e., instead of having 40 binaries, for example, 10 should be enough. The memory is common to all these elementary arithmetical organs which, in addition, can work at the same time for several different particles. Another part of the machine collates the results, performing the more involved logical computations, providing the given functions of the random steps, etc. What we allude to here is, as it were, to have almost a physical analogue element as a part of a digital computer in order to provide independent random walks or even more complicated processes. It is beyond the scope of this chapter (and its author) to propose in detail a working mechanism of this sort.

In Monte Carlo type problems we saw how independent units describing random walks and branchings (independent particles being followed simultaneously) could collate their results at the same time. I want to indicate here how certain problems of mathematical logic could be studied experimentally on computing machines by using a physical operation of projection.

It has been noted that the existential quantifier corresponds to the geometric projection parallel to one axis onto the other axis of a set located, let's say, in two dimensions. A simple case will illustrate it sufficiently: If $\phi(x,y)$ is a Boolean expression, the set of all x for which there exists a y such that $\phi(x,y)$ holds is simply the projection F_x onto the x-axis of the set F of such points (x,y) for which $\phi(x,y)$ is true. The "existential" quantifiers correspond therefore to projection. The general "for all" quantifier can be interpreted, set-theoretically, by complementation and the existential quantifier with the aid of De Morgan's rule.

If, in addition to the operations of Boolean algebra, one has a mathematical system employing quantifiers, then starting from a small number of given sets expressing the arithmetical operations, like the surfaces $z = x + y$ and $z = x - y$, one can obtain rather comprehensive mathematical theories.

The study of the algebraic properties of systems which include, in addition to the usual Boolean operations, the operations of projections and direct product is still very incomplete. Simple models of such systems were constructed by C. Everett and me. A two-dimensional projective algebra is simply a class of sets, contained in a direct product E^2 of a set E with itself, which is closed under the Boolean operations and also under the operations of projection of a set contained in E^2 onto either of the two "axes." In addition, given two subsets, $A, B,$ of the set E in our class, it is permitted to construct the set $A \times B$ in E^2. There are many problems still unsolved about such projective algebras.

Imagine now that the set E^2 (and E) is physically realized by a lattice of points on an electronic computing machine which has orders allowing it to form the logical sum of two sets and the complement of a set. In addition, and these would be essential elements, the machine should have an order which would allow the formation of the projection A_x of a set A contained in E^2 onto the x-axis, and an order for construction in E^2 of the direct product $A \times B$ of two subsets, A, B of E. It would be awkward and too time consuming, by a big factor, to have these Boolean operations and the operations of projection and direct product effected by the machine through a point-by-point construction in series. Only a device which would allow it to form from two sets A, B their sum all at once, and also at one throw the projection of a set in E^2 by a single order, would be rapid enough to permit an algebraic study of properties of such projective algebra over a finite number n of sets may be infinite, in contrast to a Boolean algebra which has at most 2^n different elements. If the space E has k points, the number of possible subsets in E^2 is 2^{k^2}, and the number of possible classes of sets is $2^{2^{k^2}}$

It is clear that the wealth of combinatorial structure in finite projective algebras is very great and, what is more, its study may provide hints for the behavior of infinite classes. Quite generally, the fear that the very large numbers appearing in combinatorial problems prevent heuristic study on computing machines is not justified. While it is true that combinatorial functions like 2^n or $n!$ increase so rapidly as to get out of the range of numbers available even on the largest computers, it should also be realized that by statistical or Monte Carlo type experiments one can obtain inequalities in enumeration problems with small samples.

Suppose $f(x)$ is a function on the set $E = 1, 2, \ldots, n$ to itself. The graph of the function $f(x)$ is then a set of points $(x, f(x))$ in the space E^2. Consider in the space E^2 the diagonal, that is to say, the set of all (x, y) such that $x = y$. The function $f(x)$ can be iterated graphically as follows: at every point $(x, f(x))$ we move parallel to the x-axis across to the diagonal, then vertically to the graph of $f(x)$, thus obtaining the ordinate x for the point $(x, f(f(x)))$. It is plausible that a single electronic arrangement should be possible whereby the function $f(f(x))$ is computed all at once for all x under consideration.

One can think of simple optical arrangements which would allow one to graph the second and higher iterates of a function by such a process. Needless to say, given two such functions $f(x)$, $g(x)$, one can obtain $g(y(x))$; it suffices to start with the point $(x, f(x))$, move to the diagonal, go to the value $g(x)$ at that point, then go back to x and "plot" the composite function One could think perhaps of a gadget involving a matrix of, say, 1000×1000 points on which such operations could be performed simultaneously to calculate composite functions.

If a clever way were found to have such additional devices joined to a computer, many combinatorial problems and some problems in analysis could be studied very rapidly. Anyone familiar with the role of recursive functions realizes that once one has the operation of composition of functions, it is possible to define binary operations or number-theoretical functions with the aid of compositions alone. Perhaps a saving in time could be obtained in a variety of problems.

CHAPTER 9

PATTERNS OF GROWTH
OF FIGURES

THIS CHAPTER CONTAINS a brief discussion of certain properties of figures in two- or three-dimensional space which are obtained by rather simple recursion relations. Starting from an initial configuration, one defines in successive "generations" additions to the existing figure, representing, as it were, a growth of the initial pattern in discrete units of time. The basic thing will be a fixed division of the plane (or space) into regular elementary figures. For example, the plane may be divided into squares or else into equilateral triangles (the space into cubes, etc.). An initial configuration will be a finite number of elements of such a subdivision and our induction rule will define successive accretions to the starting configuration.

The simplest patterns observed, for example in crystals, are periodic and the properties of such have been studied mathematically very extensively. The rules which we shall employ will lead to much more complicated and in general non-periodic structures whose properties are more difficult to establish, despite the relative simplicity of our recursion relations. The objects defined in that way seem to be, so to say, intermediate in complexity between inorganic patterns like those of crystals and the more varied intricacies of organic molecules and structures. In fact, one of our aims here is to show, by admittedly somewhat artificial examples, an enormous variety of objects which may be obtained by means of rather simple inductive definitions and to throw a sidelight on the question of how much "information" is necessary to describe the seemingly enormously elaborate structures of living objects.

Much of the work described below was performed in collaboration with Dr. J. Holladay and Robert Schrandt. We have used electronic computing machines at the Los Alamos Scientific Laboratory to produce a great number of such patterns and to survey certain properties of their morphology, both in time and space. Most of the results are empirical in nature, and so far there are very few general properties which can be obtained theoretically.

In the simplest case we have the subdivision of the infinite plane into squares. We start, in the first generation, with a finite number of squares and define now a rule of growth as follows: given a number of squares in the nth generation, the squares of the $(n + 1)$th generation will be all those which are adjacent to the nth generation square but with the following proviso: the squares which are adjacent to more than one square of the nth generation will not be taken. For example, starting with one square in the first generation one obtains the configuration of Figure 1 after five generations. See also Figure 3.

It is obvious that with this rule of growth the figure will continue increasing indefinitely. It will have the original symmetry of the initial configuration (1 square) and on the four

Figure 1.

Figure 2.

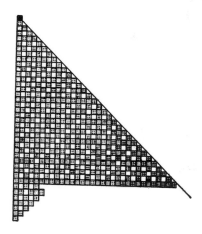

Figure 3. Starting with the black square as the first generation, each successive generation consists of those squares that are adjacent to one and only one square of previous generations. In most of these illustrations, only cells in certain directions were drawn. Growth in other directions is the same because of symmetry conditions.

perpendicular axes all the squares will be present — these are the "stems" from which side branches of variable lengths will grow.

We can consider right away a slightly modified rule of growth. Starting again with a single square and defining the $(n + 1)$th

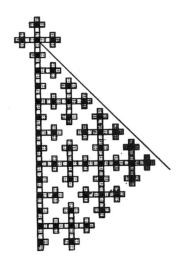

Figure 4. "Maltese" crosses. The black cells in this pattern are arranged according to the pattern for Figure 3, except that they are more spread out. Here we use the same rule as in Figure 3 with the following exception: if a cell would touch some other cell (either already grown or being considered for growth in this generation) on either a corner or a side, it is rejected. However, we made two exceptions to this restriction. (1) if the cell touches some other cell by virtue

$$\begin{array}{cc} & 2 \quad 5^* \end{array}$$

of having the same parent, and (2) in the following case: 1 2 3 4 5. The two

$$\begin{array}{cc} & 2 \quad 5^* \end{array}$$

starred elements of the fifth generation are allowed to touch potential, though previously rejected, children of the third generation. This has to be allowed to enable the growth to turn corners. Note that children of the third generation were rejected only because of the potential children of the starred members of the second generation.

generation as before to be squares adjacent to the squares of the nth generation, we modify our exclusion proviso as follows: we will not put into existence any square for the $(n + 1)$th generation if another prospective candidate for it would as much as touch at one point the square under consideration. With this second rule we obtain Figure 2 after five generations. See also Figure 4. With this rule we will again notice immediately that the "stem" will continue indefinitely, but now the density of the growing squares will be less than in the previous case. In this case again one can calculate which squares will appear in the plane and which will remain vacant.

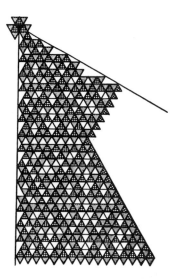

Figure 5. This pattern follows the same rule as Figure 3, except that triangles are used instead of squares.

Figure 6. This pattern follows the same rule as Figure 5 with one exception: if a new cell would touch the corner of some old cell (other than a parent), it is rejected.

A general property of systems growing under the rules (and even somewhat more general ones) is given by a theorem due to J. Holladay. At generations whose index number n is of the form $n = 2^k$, the growth *is cut off* everywhere except on the "stems," i.e., the straight lines issuing from the original point. The old side branches will terminate and the only new ones will start growing from the continuation of the stems.

One of the most interesting situations arises when the plane is divided into equilateral triangles and, starting from one initial triangle, we construct new ones, generation by generation. We can again have the analogue of the first rule; i.e., for the $(n + 1)$th generation we consider all triangles adjacent to a triangle of the nth generation. As before we shall not construct those which have two different parents in the nth generation. The system which will grow will have the six-fold symmetry of the original figure. There will appear a rather dense collection of triangles in the plane. See Figure 5 and also Figure 6. (Each figure shows a segment of the growing pattern. It represents one-half of a 60-degree section. The other half is obtained by a mirror image. The other sections are obtained by rotation.)

Another way is to take the analogue of the second rule of "conflict"; i.e., do not construct a triangle in the $(n + 1)$th generation if it would so much as touch at one point another prospective child of some other element in the nth generation. (We, of course, allow two prospective children to touch on their base from two adjacent parents.) This rule will lead to a pattern which has fewer elements and a smaller density in the plane than the one constructed under the first conflict definition. See Figure 7.

One can prove easily that the initial hexagonal symmetry will persist and that the growth will continue indefinitely with the "stems" increasing in each generation by one element, i.e., forming continuous lines. The side branches have variable lengths and get "choked off" at variable times (generations). The author did not manage to prove that there will exist infinitely long side branches. It is possible to demonstrate that there will be arbitrarily long ones.

Figure 7. This pattern follows the same rule as Figure 6 with this exception: if two new cells (other than siblings) would touch each other even on a corner, they are both rejected.

For the construction with triangles under the first rule, Holladay's cut-off property holds for generations with index of the form 2^k. Under the second rule it was not even possible to prove the value or indeed the existence of a limiting density of the triangles obtained by the construction (relative to all the triangles in the whole plane).

In the division of the plane into regular hexagons and starting with, say, again one element, one can obtain the analogues of the two patterns. See Figure 8. Again the analogue of the more liberal construction has the cut-off property. For the more stringent rule it was, so far, impossible to predict the asymptotic properties.

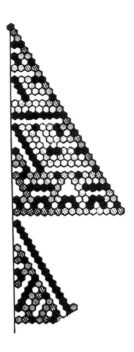

Figure 8. This pattern follows the same rule as Figure 3, except that hexagons are used instead of squares. It is disconnected because a triangle of cells is left out. This triangle is the same as that formed in the first few generations for what is drawn plus a mirror image of it.

The construction of the elements of the $(n + 1)$th generation is through a single parentage: each element attempts to generate one new one in the next generation. In the division of the plane into squares, triangles, hexagons, etc., one could adopt a different point of view: in the case of, say, triangles, one can consider instead of the areas of the triangles, their vertices only, and imagine that each pair of vertices produces a new vertex, namely the one forming the triangle with the two given vertices as their sides. Actually, the origin of the above-mentioned constructions is due to this point of view.

It is also of interest to consider problems of "binary" reaction systems. Mathematically, these involve the following situation: a great number of elements is given, each element being one of, say, three types. These elements combine in pairs and produce in the next generation another pair of elements whose types are unique functions of the types of the two parents. The problem is to determine the properties of the composition of the population as time goes on. If x, y, z denote the proportions of elements of the three types in the nth generation, then the expected value of the numbers of particles of each type in the new generation will be given by a quadratic transformation. For example, the rule could be that an x type and a y type particle together produce an x type, the $(x + x)$ a z type, $(x + z)$ a y type, $(y + y)$ an x type, $(y + z)$ a z type, and $(z + z)$ a y type. (Actually there are more than 90 possible and different such rules — we assume, however, that once a rule is chosen it is valid for all time.) The rule above would lead to the new proportions x', y', z' given as follows:

$$x' = 2xy + y^2,$$
$$y' = z^2 + 2xz,$$
$$z' = y^2 + 2yz.$$

This is a transformation of a part of the plane into itself. We have three variables, but $x + y + z = 1 = x' + y' + z'$. By iterating this transformation one obtains the expected values of the numbers of elements of each type in the subsequent generations. In the above-mentioned study some properties of the iterates of the transformation were established. In particular, in some cases there may be convergence to a stable distribution, in other cases there is a convergence to an oscillating behavior, and so on.

These studies concerned random matings (or collisions) between pairs of elements. The question arose as to the behavior of such systems if the binary production were not a random one but instead subject to some constraints, say, due to geometry. A most stringent one seemed to be to imagine, for example, that the elements form the vertices of a division of a plane into regular triangles, each vertex being of one of the three possible "colors." Then consider an initial configuration as given and assume the production of new elements by pairs of vertices forming sides of the triangular division. In the simplest case one can start with one triangle whose three vertices are all different in type. The next generation will be formed then by the three pairs as parents, and each side of the given triangle will produce a new vertex whose color is a function of the two colors of the parents. We shall obtain then a second generation and continue in this fashion. It is immediately found, however, that the construction cannot be uniquely continued. After a small number of generations it will appear that two pairs of vertices forming two sides of the configuration will have a single vertex as completing the two triangles to be constructed. Which color to assign to the new vertex? It may be that the two sets of parents will give a conflicting recipe for the color of the new point.

One way out of this dilemma would be not to consider a point for which a conflicting determination of color may be given and leave its position vacant. The patterns mentioned above can be considered as points which are of three different kinds (imagining, for example, that the new ones arise in a "molecule" as a result of a double bond, etc.) There are also other recipes for determining the color of points which were given conflicting determinations by the two pairs of parents. Rule 1 is to choose the type not involved in the conflicting determination: since there are three types, if the two determinations for the new points differ, one may choose the third one. Another rule, Rule 2, can also be considered to decide at random, with equal probability, which of the two contrasting determinations should be chosen. And Rule 3 is to choose, in case of such a conflict, a fourth color whose proportion will be denoted by w and such that an x type + w type produces x; y + w produces y; z + w produces z; and w + w produces w in subsequent combinations. This could have an interpretation of representing a molecule of a type which cannot propagate except in combination with itself. The propagation of such systems has been studied experimentally. Rule 2 in particular involves sometimes a random determination of points somewhat similar to random mating. Under all these rules there seems to be a convergence of the number of particles of different types to a steady distribution (in contrast

to the behavior given by iteration of the quadratic transformations where in many cases there is an oscillatory limit or even
more irregular ergodic asymptotic behavior). In some cases the
convergence seems to take place to a fixed point (i.e., a definite
value of x,y,z), and under rule (2) to values, numerically not
too different from the fixed point of the corresponding quadratic
transformation. It has not been possible to *prove* the existence
of a limiting distribution but the numerical work strongly indicates it. It should be noted that all the initial configurations
were of the simplest possible type, e.g., consisted of one triplet
of points.

I shall return now to the discussion of growing patterns where
we do not label the new elements by different colors but merely
consider the geometry of the growing figure. The problem arose
of considering the properties of growth of such figures with a
rule of erasure or "death" of old elements: suppose we fix an
integer k arbitrarily and to our recursive definition of construction of new elements add the rule that we erase from the pattern
all elements which are k generations old. In particular, suppose
$k = 3$ and consider the growth from squares, as in the very first
rule described on page 78, with the additional proviso that after
constructing the $(n + 1)$th generation, we shall erase all points
of the $(n - 1)$th generation. (The construction allows the configuration to grow back into points of a previous generation of
index ℓ where ℓ is less than $n - 1$.) In this construction, starting, say, with two squares, one will observe a growth of patterns,
then a splitting (due to erasures) and then later recombinations
of the pattern. A search was undertaken for initial patterns which
in future generations split into figures similar or identical with
previous ones, i.e., a reproduction, at least for certain values
of the index of generation. It was not possible, in general, even
in the cases where a growth pattern without erasure could be
predicted, to describe the appearance of the apparently moving
figures, which, in general, exhibit a very chaotic behavior. In
one starting configuration, however, one could predict the future
behavior. This configuration consists of two squares touching

each other at one point and located diagonally. Under rule (1) with erasure of the third oldest generation, this pattern is reproduced as four copies of itself in every 2^pth generation ($p = 1,2,3,...$), displaced by 2^p units from the original pattern. The same behavior holds for starting patterns of, say, four squares located diagonally, or 8 points, or 16 points, and so on.

In case of a triangular subdivision the behavior of growth with a rule of erasure for old elements was also experimentally investigated. The process of growth was considered as follows: given a finite collection of vertices of the triangular subdivision of the plane — some labeled with the index $n + 1$ and others with n, one constructs the points of the $(n + 1)$th generation by adding vertices of the triangles whose sides are labeled either with $n - 1$ and n or n and n, again, however, *not* putting in points which are *doubly* determined. One then erases all points with the index $n - 1$. In case of squares our rules of growth enable the pattern to exist indefinitely, starting with any nontrivial initial condition. This is not always the case for triangles. In particular a starting pattern of two vertices with the same generation terminates after ten generations — that is to say, all possible points of growth are conflicting ones and these are not allowed by our rule of construction. One has to point out here that in the case of the "death" rule which operates by erasure of all elements that are k generations old, the initial configuration has to specify which elements are of the first and which of the second generation. Two vertices, one labeled first and the other second generation, will give rise to a viable pattern.

In three-dimensional space a similar experimental study was made of growth of patterns on a cubical lattice. The rules of growth can be considered in a similar way to the recipes used in two dimensions. Starting with one cube one may construct new ones which are adjacent to it (have a face in common). Again one will not put in new cubes if they have a face in common with more than one cube of the previous generation. The analogue of the first rule gives a system whose density in space tends to 0. This is in contrast to the situation in the plane where a finite density was obtained for this case.

R. Schrandt has investigated on a computer the growth of a system with a rule for erasure of old elements. The case of erasure of elements three generations old was followed. The patterns which appear seem to be characterized by bunches of cubes forming flat groups. These groups are connected by thin threads.*

These heuristic studies, already in two dimensions, show that the variety of patterns is too great to allow simple characterizations. I have attempted to make corresponding definitions in one dimension with the hope that some general properties of sequences defined by analogous recursive rules would be gleaned from them. Suppose we define a sequence of integers as follows: starting with the integers 1,2 we construct new ones in sequence by considering sums of two previously defined integers but not including in our collection those integers which can be obtained as a sum of previous ones in more than one way. We never add an integer to itself. The sequence which starts with 1 and 2 will continue as follows: 1, 2, 3, 4, 6, 8, 11, 13, 16, 18, 26, 28,.... The integer 5 is not in it because it is a sum of two previous ones in two different ways. The next integer which is expressed in one and only one way of the sum of previous ones is 6; 7 has a double representation but 8 is uniquely determined. 11 is the next and so on. Starting with 1 and 3 one obtains the following sequence: 1, 3, 4, 5, 6, 8, 10, 12, 17, 21,.... Unfortunately, it appears that even here it is not easy to establish properties of these "unique sum sequences." For example, the question of whether there will be infinitely many twins, i.e., integers in succession differing by two, seems difficult to answer. Even a good estimate of density of these sequences relative to the set of all integers is not easily made.

The aim in presenting these disconnected empirical studies was to point out problems attending the combinatorics of systems which, in an extremely simplified and schematic way, show a growth of figures subject to simple geometrical constraints. It

---

* See chapter 10.

seems obvious that, before one can obtain some general properties a great deal of experimental data have to be surveyed.
The work is continuing and perhaps some more general properties of their morphology will be demonstrable.

CHAPTER 10

MORE ON PATTERNS OF GROWTH

with Robert G. Schrandt

IN THIS CHAPTER we shall discuss briefly some empirical results obtained by experiments on computers. We continue the work described in the previous chapter.

We shall describe patterns "growing" according to certain recursive rules. This growth takes place on the grid of squares in a plane. We give an example of such a pattern growing on the infinite plane and then discuss patterns of growth in an infinite strip of a given width where a periodic growth is observed. We discuss the behavior of figures growing according to our rules, with a new proviso: every element of the figure which is older than a certain specified number of generations, say, two or three, "dies" (i.e., is erased). This makes the figure move in the plane. We show some cases of such motion, with occasional splitting of the figures into separate connected pieces. In some cases these figures are similar to the original ones and so we have phenomena both of motion and of self-reproduction.

As another amusement we tried out on the computer the following game: starting in the plane with two separate initial elements, we let them each grow according to our rule (including erasure or death of the "old" pieces); then when the two patterns approach each other we still apply the rule of a further growth of each of the figures with the proviso that the would-be new pieces are not put in if they try to occupy the same square. This gives rise to a game for survival or a "fight" between two such systems — in some cases both figures die out.

Finally, we give an example of a similar process of growth in three-dimensional space (subdivided into regular cubes) with one of our rules for recursive addition of new elements. We enclose pictures of such an object grown to the 30th generation, of which we made a model and photographed it from two angles (Figure 1). The model represents the part grown in only one octant of space. In each octant there is a further three-fold symmetry and we show only *one* of the three parts. The cube on the extreme left in the upper picture is the start of the pattern. The dark cubes represent the 30th generation.

Our examples show both the complexity and the richness of forms obtained from starting with a simple geometrical element (a square or a cube!) and applying a simple recursive rule. The amount of "information" contained in these objects is therefore quite small, despite their apparent complexity and unpredictability.

If one wanted to define a process of growth which operates continuously rather than by discrete steps, the formulation would have to involve functional equations concerning partial derivatives.

It appears to us that the geometry of objects defined by recursions and iterative procedures deserves a general study — it produces a variety of sets different from those defined by explicit algebraic or analytic expressions or by the usual differential equations.

Our growth is in the plane subdivided into regular squares. The starting configuration may be an arbitrary set of (closed) squares. The growth proceeds by generations in discrete intervals of time. Only the squares of the last generation are "alive" and able to give rise to new squares. Given the nth generation, we define the $(n + 1)$th as follows. A square of the next generation is formed if

    (1) it is contiguous to one and only one square of the current generation;

    (2) it touches no other previously occupied square except if the square should be its "grandparent";

Figure 1. Two views of a three-dimensional object grown from a single cube to the 30th generation.

(3) of this set of prospective squares of this $(n + 1)$ generation satisfying the previous condition, we eliminate all those that would touch each other. Again we make an exception for those squares that have the same parent; these are allowed to touch.

In three dimensions the growth rule which we have adopted is the same. One merely replaces the squares by cubes and observes our three provisions.

We present examples of this type of pattern in Figure 2a and b. Our starting configuration was always a single square. The patterns which are shown are plotted only in one quadrant of the plane and they show the result of 100 and 120 generations of growth respectively. The second picture is shown only on a large square with 100 units on the side and the portion of growth that extends beyond 100 units horizontally or vertically is not plotted. Our figure is symmetric about the diagonal of the square. The density of the occupied squares is about .44. There is no apparent periodicity in portions of this pattern. As shown in the preceding chapter the "stems" grow indefinitely on the sides of the quadrant. The side branches split off from the stem. It is not known whether some of these side branches will grow to infinite length or whether they will all in turn be choked off by side branches growing from the stem at later and later times.

In Figure 3 we show a pattern which grows from an initial configuration, consisting of three squares that are not continuous. We have chosen three squares at vertices of an approximately equilateral triangle. One will note that the patterns in the subquadrant are identical to those of Figure 2. There are borders or strips between them because of interference between patterns generated by the individual starting squares. There are borders or strips between the first and second and between the second and third quadrants. In the other two cases these borders reduce to stems, since two of the starting squares generate patterns symmetric with respect to a 45° line through the center of the triangle.

One can restrict in advance the growth of a pattern to an infinite strip of finite width in the plane. One then observes a

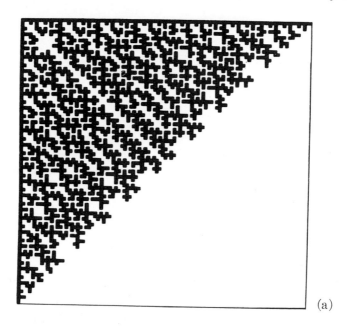

(a)

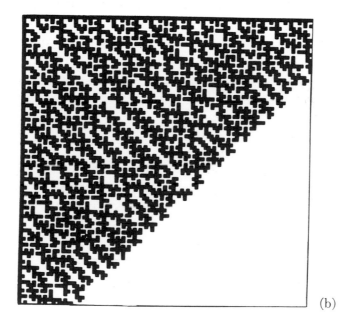

(b)

Figure 2. Pattern grown from a single square. (a) After 100 generations; (b) After 120 generations.

Figure 3. Pattern grown from three noncontiguous squares.

periodic growth. The proof that in a finite strip the pattern must be periodic is easily obtained. On inspection of our growth procedure one observes that the last generation is confined to a part of the strip which extends through its width and a similar distance in length back of the most forward square. There are only a finite number of possible patterns in such a square. Therefore, a configuration must repeat itself and from then on the whole process will start all over again.

Figure 4 shows different patterns generated in strips of widths from 8 to 15 through 100 generations of growth. In each case the starting point is a single square in the upper left-hand corner of the strip. In Table 1 we give the observed lengths of the periods, for strips of various widths (1–17). There seems to be no simple relation between the width of the strip and the length of the period.

We have experimented with a rule for erasing, that is to say, the elimination of a part of the pattern after it is a fixed number of generations old. For this we have adopted a simpler rule of

96

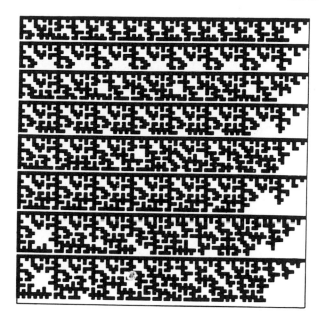

Figure 4. Patterns generated after 100 generations in strips of width from 8 to 15.

Table 1

Width of Strip	Period of Pattern
1	1
2	2
3	3
4	5
5	5
6	8
7	13
8	13
9	13
10	26
11	13
12	91
13	13
14	106
15	106
16	75
17	93

growth of the pattern, assuming only properties (1) and not the additional (2) and (3). Each square that is a certain fixed number k of generations old is erased or "dies" and becomes unoccupied. (Later on, the pattern may grow back into these unoccupied positions.) For example if $k = 2$, we grow the squares of the $(n + 1)$th generation from those of the nth and then erase those of the $(n - 1)$th.

For $k = 2$ or 3 the pattern will move, and it may split up into disconnected pieces, as in Figure 5a. It turns out that certain parts of it replicate themselves in shape and these repeat as subpatterns. One such consists of a straight strip of squares with two additional squares on each end. This is rather frequent and we call this replicating subpattern a "dog bone" (see Figure 5b).

We have experimented also with our "death rule" keeping three generations alive.

Another construction which we have studied concerns the behavior of such patterns in a finite portion of the plane. We have adopted a large square as the space for growth. Its boundary acted as an absorber so that each square which would possibly grow from a square on the boundary was not considered. This was studied for $k = 1$; that is, given a configuration, we produce the next one and immediately erase the starting one. Starting with an initial configuration the pattern will grow and either eventually "die," that is to say, disappear altogether (under our "death rule"), or else the pattern will become periodic in time and continue in this fashion indefinitely. It seems that in "most" cases the pattern eventually disappears.

We have run the problem on a computer for many cases. This we have done in four different sizes of the large square in which our game takes place. A sampling was obtained for sizes of the large square from 2×2 up to 8×8. As an example, consider the square of size 6×6. There are, of course, 2^{36} possible initial configurations. Out of these we have chosen 132 such configurations at random (specifying that each of the 36 squares has ½ chance of being occupied initially). Each of these 132 different starting configurations grew until it became periodic, or else degenerated to zero. Let ℓ be the length of the period of a

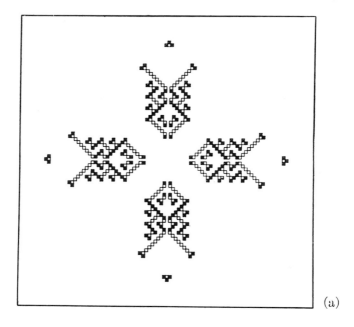

(a)

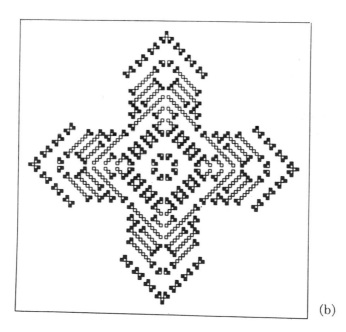

(b)

Figure 5. Patterns grown with an erasure rule.

sequence of states, with $\ell = 1$ for a terminating sequence. The values of ℓ which were observed were 1, 4, 6, 8, 12, 24; 87 of our 132 starting positions had $\ell = 1$ since they had degenerated to zero. Let s be the number of time steps in a terminating sequence of states. The values of s ranged from 11 to 109, with an average value of the length of the sequence in our sample equal to 33. We also tried 15 random starting configurations chosen in an 8×8 square. Ten of these lead to terminating sequences. The values of s ranged from 49 to 397, with ℓ values of 1, 8, 12, and 16. Ten of our 15 experiments terminated in zero.

For the case of $k = 1$ (keeping only one generation), the status of any square in the generation $(n + 1)$ is determined by the state of its four neighboring squares in the nth generation in a very simple way. Let us assign a 1 to an occupied square and a 0 to an empty one. Let us use the two operators $(\cdot)$ and $(+)$; we use the $(+)$ modulo 2; that is, $1 + 1 = 0$. If a_n, b_n, c_n, d_n are the four neighbors of a point x_n and all the symbols have values 1 or 0, then the state of the point x in the next generation is simply

$$x_{n+1} = \bar{c}_n \cdot \bar{d}_n \cdot (a_n + b_n) + \bar{a}_n \cdot \bar{b}_n \cdot (c_n + d_n),$$

where the bars above the symbols represent the complement (also modulo 2).

If the whole region in which the game is played is bounded by a large square, we will assume that the values on the boundary are always 0. The state of configuration at time $n + 1$ is obtainable by a fixed transformation from the state at time n.

One of the first things to determine is the existence of the states which are self-replicating; that is, they reproduce themselves immediately. These are the fixed points of the transformation defined above. It is easily verified that there are none such except those that are identically 0 for squares of the size 2×2 and 3×3. There exists just one such state for the 4×4 square. This is given by

$$\begin{vmatrix} 0 & 1 & 1 & 0 \\ 1 & 0 & 0 & 1 \\ 1 & 0 & 0 & 1 \\ 0 & 1 & 1 & 0 \end{vmatrix}$$

For the 5 × 5 square there are these two:

$$\begin{vmatrix} 1 & 0 & 1 & 0 & 1 \\ 1 & 0 & 1 & 0 & 1 \\ 0 & 0 & 0 & 0 & 0 \\ 1 & 0 & 1 & 0 & 1 \\ 1 & 0 & 1 & 0 & 1 \end{vmatrix} \quad \text{and} \quad \begin{vmatrix} 1 & 1 & 0 & 1 & 1 \\ 0 & 0 & 0 & 0 & 0 \\ 1 & 1 & 0 & 1 & 1 \\ 0 & 0 & 0 & 0 & 0 \\ 1 & 1 & 0 & 1 & 1 \end{vmatrix}.$$

There are none for the 6 × 6 case. Here is an example of a 17 × 17 matrix. Let A be the second of the two 5 × 5 matrices. Let N_C and N_R be 5 × 1 and 1 × 5 matrices respectively with zero elements. Then the matrix

$$\begin{vmatrix} A & N_C & A & N_C & A \\ N_R & 0 & N_R & 0 & N_R \\ A & N_C & A & N_C & A \\ N_R & 0 & N_R & 0 & N_R \\ A & N_C & A & N_C & A \end{vmatrix}$$

is self-replicating.

We may start, on a large finite square, with two different initial configurations each labeled by a different color to distinguish one set from the other. We can let them each grow according to our rules and then when they approach each other, we may apply our rule to both of these sets taken together. The growth of these patterns then is subject to our restrictions for elements of the new generation within themselves separately, and when they are almost in contact to the two taken together. One or both of these systems may then go to zero or else one may survive, for some time or indefinitely.

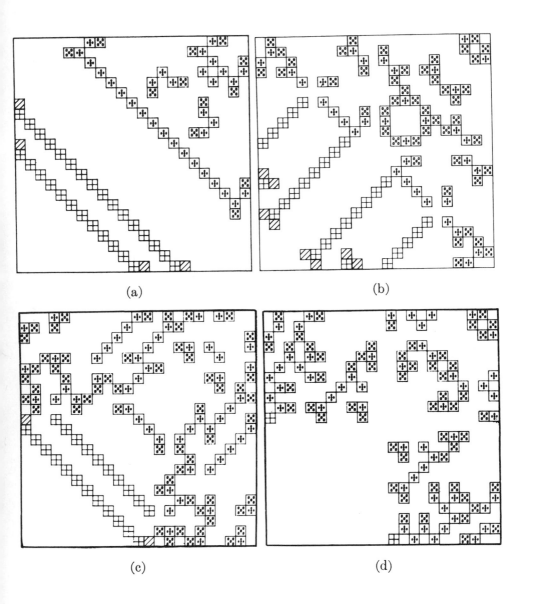

Figure 6. A "fight" between two patterns.
(a) The situation at generation 16; (b) The situation at generation 25;
(c) The situation at generation 32; (d) The situation at generation 33.

Figure 6 illustrates one case of such a "fight" between two starting patterns in a 23 × 23 square. It shows the situation at generations 16, 25, 32, and 33 respectively. In this case we kept two generations before erasure (i.e., $k = 2$). The pattern in the lower left-hand section of Figure 6a is called pattern A; the squares with diagonals belong to the present generation. The pattern in the upper right-hand section of Figure 6a is called pattern B; the squares with five small squares inside belong to the present generation. We assumed as initial conditions for pattern A one square in the extreme lower left-hand corner, and pattern B started from one square placed one unit of distance off from the upper right-hand corner. After 33 generations pattern B won; that is, at that time A was completely erased.

In another game we started with two single squares in the same relative positions from corners. We had, however, pattern A with the rule of growth involving keeping only one generation before erasures; pattern B was growing with two generations before the "death" of the old pieces. For this case pattern A won in 112 generations on the 23 × 23 board.

CHAPTER 11

HOW TO FORMULATE MATHEMATICALLY THE PROBLEMS OF THE RATE OF EVOLUTION

WHEN I WAS about 14 years old, I was very much interested in mathematics and wanted to become a mathematician, but I read a lot of other things, too. Conan Doyle's story, "The Lost World," was a fascinating book. The main character, Professor Challenger, was a natural scientist. Those who have read the book remember, perhaps, his terrific temper and the violent debates in which biologists engaged at that time and how Challenger's terrible temper made such scientific meetings exciting. Then, at the age of 18 I attended a mathematical meeting and was terribly disappointed that nothing of the sort occurred. It was all too calm and quiet.

I believe that most mathematicians and perhaps lay people, in general, tend to be somewhat congenitally neo-Lamarckians. Some years ago Victor Weisskopf and I had a discussion. We wondered how it appeared extremely unlikely *a priori* that in the short span of one billion years, all the wonderful things that we see now could have appeared, due to successive random mutations. It seemed to require many thousands, perhaps millions, of successive mutations to produce even the easiest complexities we see in life now. It appears, naïvely at least, that no matter how large the probability of a single mutation is, should it be even as great as one-half, this probability would be raised to a millionth power, which is so very close to zero that the chances of such a chain seem to be practically nonexistent.

A mathematical treatment of evolution, if it is to be formulated at all, no matter how crudely, must include the mechanism of the advantages that single mutations bring about and the process of how these advantages, no matter how slight, serve to sieve out parts of the population, which then get additional advantages. It is the process of selection which might produce the more complicated organisms that exist today.

I have done some very schematic thinking on the mathematics of such a process and I want to make some remarks which are surely not correct in a realistic sense but might be relevant for the approach to a quasi-mathematical discussion. These philosophical and general methodological remarks could form a basis for what could be mathematized in the future. What I will do is, as it were, pick out various items and try to show how mathematical schemata could be formulated.

This reminds me of a story about two persons who collaborated on writing musical comedies. One of them was writing the libretto, the other one the music. One day they got together and one asked the other: "Do you have any ideas about the music?"

"No, do you have any ideas about the libretto?"

"No."

"O.K., let's go on and write the play."

Perhaps this is the situation now in attempting to make something like mathematical treatments of biological problems.

It seems to me, nevertheless, even if it is a little premature, that it is still a good exercise to think about these things. Perhaps in some years something amusing or even useful may come out of it.

Before any such thing can be treated mathematically one must have numerical values for certain quantities or parameters, which to a mathematician are just parameters or only letters; but, if one wants to use it in any biological illustration, one must have their given numerical values.

I will give you a whole set of such parameters with values

that are important to know. The trouble is that, at present, realistic definitions of these parameters, not to mention the numerical values, are completely unknown.

First of all, let us start with a total number of some extremely primitive organisms, perhaps some simple bacteria or pre-bacteria which existed at one time, formed by some random combinations through a chemical process. If one wants to make it easier for the Darwinian Theory, this number should be taken as large as possible. N, then, is a number of some very primitive individuals which existed, say, a billion years ago.

The other parameter, which is obviously of great importance, is the total time, T, that life has existed so far. This is, let us say, one billion years. All these "parameters" are uncertain by factors of hundreds, thousands or more. This last one is perhaps better estimable than the others.

One needs also a number, τ, the lifetime of a single individual. Again, for our purpose, it is "better" to make it as short as possible. We want to have as many generations as possible, since mutations coming between them, then, are transmitted. Assume that it is just one day, for example. Now comes the only number which is easily computable on the basis of the other numbers, that is the number of generations that have existed. This is simply T/τ.

Now come numbers really difficult to estimate — first of all, the frequency of mutations per lifetime of one individual. These mutations might be due to cosmic rays or chemical mutagens. We can call their frequency α. We want, however, not the frequency of all mutations but the frequency of the *favorable* ones, whatever that means. This number, even in some sense averaged as one per individual (and there are different individuals, of course), is very hard to estimate. It is certainly very small. α is thus the probability that there will be a favorable mutation in the genetic makeup of one individual in his lifetime, in any locus, any cell, any gene. I will try to define what I mean by "favorable" for the purpose of one very crude and oversimplified mathematical scheme. I have to warn you, should it be necessary, not to put any credence whatsoever on the value of these parameters. In some future mathematical theory it will, however, be necessary to get numbers for such values. It seems to me that the final probabilities, even if we assume some values from the present guesses, will not be fantastically small. We will come to that later.

Let us now assume that α is a small number, say 10^{-10}. Certainly as you look at the frequency of the radiation impinging on a small object during a short lifetime, one obtains extremely small values.

Now we come to another number perhaps more difficult to guess or to estimate. Call it γ. It is supposed to express the differential advantage of one "favorable" mutation. How do we express it mathematically? Suppose that an individual received, by chance, a small "improvement" in his genetic makeup. How does one translate its value for the branching process of the replicating chain of living descendants? One can do it in many different ways. One way, purely mathematical, is to say that an individual which possesses this one tiny "improvement" relative to the other living organisms has a slightly better chance to survive; or, to formulate it more easily mathematically, it has a slightly higher number of offspring than the other. Please note that we speak of one small mutation. If one wants to produce some organ like an eye, together with the whole apparatus of the nervous system connecting it with what we might call a "brain," one seems to need an enormous number of such successive steps. Let me stress one very important *caveat*. This point of view of a great number of successive steps is not really absolutely necessary. One might think of another process, involving operations with more abstract changes or "rules" than the mere accumulation in succession of single pieces of "hardware".

But, naïvely, if one knows nothing about the actual world or even about biology, one might say a single one of a million successive improvements, which culminate in something like an eye, does not confer any noticeable advantage. It is really very infinitesimal. Obtaining one improvement, which gives, say, a slight photosensitivity, gives a minimally greater chance for the individual to reproduce itself. Let us say mathematically that, instead of having just one offspring, one will have $1 + 10^{-6}$, or something of this order, offspring.

In the first problem which I will mention we have only mitosis. A single bacteria reproduces itself or several bacteria like itself. In the first problem there is no question of sex. I have tried to play games on a computer with such a system. Of course we chose different and manageable values of our parameters to make it practical in the sense that the calculations could be done in a reasonable time. This first problem, called "Adam," refers to unisexual reproduction. The second problem studied, called "Eve," involved a mechanism of reproduction which is "better" — in the sense that the offspring can inherit some good characteristics of the parents independently from both of them and does not have to engage in the hopeless game of waiting for a mutation to come. The third problem, which we called "Lillith," deals with the assumption that the mating between individuals will not be assumed to be completely random as in "Eve," but the phenotypes are able to recognize similar ones, to some extent, and mate with each other, producing better chances of more "improvements" for the offspring.

Note that γ in this scheme has the interpretation that an individual which has one extra improvement has $1 + \gamma$ descendants instead of 1. We now need another parameter and its value. Somebody has to tell the poor mathematician *how many* successive improvements are necessary to produce something complicated, such as the visual system. This is a great number, no doubt, since random mutations due to, say, cosmic rays seem to be in the nature of changing the value of one bit in a code. In a random process this reproduces just one change, mostly nonsensical, but occasionally will have the effect of one rather small addition or change in the resulting apparatus. Even if one could write down a description, so complete that the computing machine could utilize it to formulate the "construction" of such an object, the number of words in this description must be very large, perhaps a million. Certainly within several or many powers of 10 these are unknown. However, let us now try to make a simple, really trivial, calculation by orders of magnitude. Let us assume that we have 10^{11} individuals. Each lives one day and let us say that the chance of a favorable mutation or "improvement" is 10^{-10} per individual. This way, on the average, there will be ten individuals in the population that have received it per generation. Now, how long does one have to wait before the bulk of the population will receive this "improvement"? It depends on γ. γ is, of course, a small number and with a tiny advantage given by this one improvement, let us say γ is 10^{-6}, it will take about 10^6 generations before most of the population is endowed with this improvement.

A biologically naïve mathematician like myself would reason as follows about γ: If you are going to produce a complicated object, an eye, by many successive changes, any one of these gives only a tiny little bit of advantage for survival, if indeed any at all.

In 10^6 generations, most of the population will have this advantage even with this small value of γ. In 10^7 generations, almost all individuals will have it. Now, remember that we want not just one improvement but 10^6 in succession. So we need about 10^{13} generations, not much more. There will be only ten individuals which received an improvement in one generation in the specific loci which determine the eye.

Now there is a very nice and simple mathematical technique for describing processes starting with a single object, which then duplicates and gives 0, 2, 3 or more descendants. It is called the theory of branching processes. It deals with asexual reproduction and gives methods to calculate the number of existing particles of various kinds, in future generations; and other questions of this sort. I would like to stress that a corresponding theory for branching with "sex," where particles get together at random and then produce offspring, i.e., a combination of a binary process of mating and of reproduction, is mathematically much more difficult, and no exact theory exists as yet. The first theory is linear, to use the technical term — at least the expected values for the numbers of particles of various types, at various times, are obtained by repetition of a linear operator. This, mathematicians know quite well how to handle. When we have two sexes and the process involves random pairings and reproduction, even the expected values of the fundamental quantities become quadratic functions. If one iterates such quadratic operators, this becomes very complicated and only the crudest beginning of a mathematical theory exists so far.

If we have two types of particles, one has the following situation. First we assume a random pairing. Then each pair, just to simplify it mathematically, produces exactly two offspring. We will allow some pairs to produce a little more than two, on the average, let us say, $2 + \gamma$, where γ is a small number, i.e. a small fraction of the population will produce three instead of two. The value of the small fraction is determined by γ. Then the process repeats again and to illustrate it by a graph instead of a "tree," which gave a schematic picture of a nonsexual process by mitosis alone, we will have a combination of tree-like growth with loops which show which pairs mated. A continuation of such may be called, perhaps, a "pear tree." It is amusing to try to develop, purely mathematically, a little theory for this process. It is much more difficult than for simple trees.

Coming back to the "biology", one sees, and natural scientists have realized it for a very long time, of course, that the situation is more favorable. The offspring may inherit, independently, improvements from both parents. Suppose one parent has ten improvements and the other one has twelve improvements, perhaps different ones. Then there is a chance for the offspring to have as many as twenty-two improvements. This chance is quite small, if we assume that each improvement has a 50% chance of being inherited by the child. However, the chance of having fourteen instead of eleven, which is the average, is not very small and, even though the expected value is just the average, even small fluctuations have a good chance of producing fourteen, as we said.

This is much more favorable than waiting many generations for a random improvement coming through mutation. In this scheme then, the fluctuations, together with the survival of the

fittest, lead to a much faster acquisition of "favorable" mutations than the pure mitosis problem. We cannot deal with it
by calculating the expected values or means alone. It is the fluctuations, and the mechanism of greater fertility of individuals
with more favorable markers, which lead to the acceleration of
evolution.

What about the question of how to treat mathematically the
advantage of having n more favorable markers than the average
population? We treat it by simply using $n \times \gamma$ for the chances
of more offspring. In other words, the relative advantage to an
individual of having n more improvements than the average
number in the population is a chance $n \times \gamma$ of having an extra
child. At first it seems that one cannot really use this recipe
because if n is very large, the number of offspring would become
nonsensically large compared to the rest of the population. The
point is, however, that in the actual processes calculated, n never
exceeds a very modest value, ten or so.

To describe quite completely or rigorously the scheme that
we tried on computers would take a great deal of time. It is essential for following the population on a computer, or even in any
mathematical treatment, to "normalize" the population in such
a way that the total number of individuals never surpasses a
fixed given value. A supercritical system would have the number
of individuals tending to infinity. In reality the globe is a finite
space and the total amount of food available is also finite, so
in dealing mathematically with such an expanded scheme, one
has to cut down periodically the total population to a fixed
number of individuals. This is one of the reasons why an analytic
formulation of the problem, using differential-integral equations,
is very difficult and would be unrealistic. A numerical investigation is, therefore, not only the thing to do in practice — the
problem really is a finite one.

The result of the calculation is that in a population with two
sexes, the average number of improvements as a function of
time or generation number increases rapidly — it is much better
than the purely linear process of one sex. The great rapidity
of the process of acquiring new favorable mutations in the prob-

lem of two sexes is due to fluctuations by which the offspring acquires more than the average of the two sets from the parents. The problem has to be considered in such a way that the population is periodically normalized to a fixed number due to the finite amount of food available, also that the relative advantage of an individual with n improvements over the average in the population is a linear function of n. It is these two assumptions together that make an analytic formulation of the problem difficult, and the behavior for a finite population cannot be imitated by a mechanical scheme dealing with a potentially infinite population.

The average is always increased. The growth is quadratic or faster than quadratic, i.e. the number of improvements in the average of the population grows that rapidly, being a function of the number of generations, i.e. time.

Sometime, of course, one will have to try to put some realistic parameters in such models, and here is a great problem. One has to interpret the nature of genes or the action of *some* genes in a somewhat more abstract way than has been done heretofore, either in general discussions or in models.

Now, a group of *things* or a group of *rules* behaves differently from the mere sum total of the individual components. In our situation it means perhaps that the action of a collection of some genes is not merely a sum total of their individual actions but an action of what mathematicians might call variables of higher type. It is the *recipes* or *rules* of coded sentences which do more than the individual actions added together. There are probably genes whose nature is to code or to program the actions of others. There may be even genes controlling, logically, still higher schemes: i.e., operating on classes of functions, etc. It is entirely possible that by some recursive rules the very prescriptions for action follow, reacting to the factors of the external world, producing from a small number of symbols or objects an ever growing variety and complexity of new prescriptions. Therefore, it is not clear at present how many bits of information are really needed to define even very complicated entities like ourselves. Perhaps this number is rather moderate or even small.

Let us return to the patterns of growth of figures. We have seen in the preceding chapters that they are defined by a simple recursive rule, but in the process of growing assume unbelievably complicated shapes later without intervention of new "mutations." If this can be formalized or pursued on more "mechanistic" models it might be that the whole discussion on whether there was enough time for evolution and selection to work would be premature or irrelevant or wrong. However, I tend to believe that these first attempts to formulate mathematically such simplified models of evolution do have some value.

Of course, the game that nature plays is really much more complicated, even from the purely logical point of view. It seems to be a game between different branching processes, each going to infinity. The question is how to formulate, without going into teleology or metaphysics, these more functional or abstract properties of genes. No doubt such formulations will come. To a layman like myself, the discoveries in molecular biology of the last fifteen years, following from the discovery of the idea and the mechanism of the genetic code, are extremely impressive. One still does not understand, as far as I know, the program for translating the code into action: why and how the transfer RNA goes to different places, does what it is supposed to do, and how the course of *action* follows from the code, which, *prima facie,* seems to dictate only recipes for mechanical or chemical actions. All this must mean that some codes have more abstract interpretations than merely rules for chemical combinations. I am not saying that in every cell there exists an analogue of a brain which operates with instructions, but undoubtedly the code must contain some elements which deal with rules for construction and *organization* and, therefore, are different from the lowest level of recipes.

CHAPTER 12

SOME FURTHER IDEAS AND PROSPECTS IN BIOMATHEMATICS

IN THIS CHAPTER, which is intended as a partial preview to what follows, we select only a few ideas of more general conceptual character in the mathematical treatment of problems arising in biology. The choice of topics is dictated largely by my predilections and in no way shall we attempt to illustrate the vast and rapidly increasing field of applications of mathematical techniques or even, more generally, the mathematical mode of thinking in the enormous variety of problems in biological studies. Indeed, to present anything like a comprehensive picture would require bulky and manifold volumes of very heterogeneous contributions. Towards the end of this chapter, however, we shall list a number of biological fields where mathematical techniques have been employed.

The role of mathematics in theoretical physics, in astronomy, and perhaps to a smaller but increasing extent in chemistry, is obvious and generally understood. These so-called exact sciences have been mathematized to such an extent that, in theoretical physics for example, ideas that cannot be *stated* mathematically are not considered to belong to the actual body of a physical theory. As is well known, the ideas of mathematics have themselves been to a very large extent stimulated and influenced by the problems that astronomy and physics have posed. The abstractions from situations in the "real world" have formed the motivations and justifications for logical or mathematical axioms and postulates underlying the fundamental constructs of mathematics. The question to what extent "pure thought" itself has, conversely, influenced the formulation of the "real phenomena" and whether one can indeed sharply separate the two [i.e., (a) the external world suggesting or, even more strongly, forcing our modes of thinking; (b) conversely, the established patterns of our logic and mathematical thinking leading to selection of our ways to describe physical reality] constitutes a philosophical question which we shall not, of course, go into here. In recent decades and years there appears an increasing willingness on the part of biologists to attempt very general points of view and "universal" schemata even though "a few exceptions" might be allowed. There is a growing likelihood that many areas of the study of the living processes may be mathematized and that some general schemata will be useful for an efficient and searching summary of new facts, which may lead to economy and success in the search for "predictions." This has happened in physical sciences in celebrated instances: for example, the abstract theory of groups was found useful not only in efficiently ordering the variety of spectral lines as a rational filing system but also in a more fundamental way in leading to the prediction of the existence of new types of particles. The abstract theory of infinitely dimensional spaces, Hilbert space especially, has been of fundamental importance in the mathematics of quantum theory. In both cases the mathematical ideas were developed before the applications. (It is a coincidence that the word "spectrum" for the collection of lines in the emission of light is the same as "spectrum" of a linear transformation in Hilbert space, a term coined by mathematicians.)

Will some similar great discoveries or theories in biology be found and formulated on the basis of abstruse mathematical theories? Will some experimental discoveries be hastened by mathematical deductions from general principles? The non-Euclidean geometry had its origin in the purely theoretical speculations of mathematicians. After Riemann, in the hands of Einstein, it turned out to be the proper mathematical description of the physical reality of the universe of masses in the large. It generalizes the Euclidean geometry found in our immediate environment. Is one allowed to speculate perhaps that some very different forms of life might exist in stellar systems other than our own? That they may still embody some general properties of functions of what we call life, and may yet not be based on all the *special* processes of chemistry that seem universal or exclusive on our earth?

In the arrangements and the properties of the nervous system and in the brain itself, do certain characteristics that are more general lie under a number of more special properties? Could one conceive of forms of functioning of the "intelligence" other than our own mechanisms, forms that would involve different types of thought processes? Questions of this sort now belong to philosophy, or perhaps even to science fiction, but one should not ignore their existence among the rapidly increasing specialization of technical mathematical descriptions of the diverse processes occuring in living matter.

The kind of mathematics that will be useful in the description of the patterns and processes of living matter and of its components is hard to delineate. Classical analysis, i.e., the differential and integral calculus, and differential and integral equations, provides the tools for description of the physical and chemical processes. These can be formulated once the quantities involved are assumed to be linked by relations that are verifiably postulated. So, for example, the flow of blood in the vessels or other liquids through channels of given physical characteristics can be described by functional relations, and the methods of classical analysis can be used in an attempt to solve questions concerning the behavior of the circulation, given the driving pressures, viscosities, etc. Problems involving rates of growth and the rates of chemical reactions can be described by (in general nonlinear) differential equations.

Other mathematical methods traditionally of use in natural sciences are the theory of probabilities and the application of statistical methods in general. A very large part of such applications are in the study of *sets* of organisms, i.e., population theories and the field of genetics. In many of these problems the intrinsic complexity of the data is such that explicit solutions (including forms) of the problems are not to be expected and numerical work is necessary. It is here that the development of the fast computing machines is of the greatest importance and value.

A field of mathematics loosely called combinatorics increasingly shows promise in the attempts to formulate the "schemes of things" in biological problems. It is hard to delineate satisfactorily the scope of this field. Its elementary, more classical, and older parts are often defined as the study of permutations and combinations of objects. Nowadays one could include in it the study of *patterns* and of *processes* involving discrete classes of sets. Later on we shall give several examples of problems and conceptual setups of biological situations in such a framework. Clearly, there isn't any sharp division between various mathematical fields and methods used in biological situations. As an example, we might mention the great role of classical mathematical analysis, the theory of discrete groups, the theory of Fourier transforms and their inverses which, together with the use of electronic computers, enabled Kendrew, Perutz, and others to discern the spatial structure of certain proteins. In this problem a confluence of mathematical techniques, experimental methods of X-ray physics, and electronic technology led to one of the triumphs of modern science.

One can observe the complementary roles of discrete and continuum methods in applications of mathematics. The distinction is, so far as problems in this chapter are concerned, principally one of methodology and outlook. If in the world of the natural sciences, living matter may be considered to be composed of atoms that form a discrete substratum of the phenomena we want to describe and study, the method of the continuum, mathematically using classical analysis (i.e., differentiation and integration), forms a *convenient* idealization and a most efficient *algorithm* to deal with the very great number of perhaps individually separated, small quantities. Needless to say, in the description of the growth of organs, in the action of body fluids, etc, and even in the study of very large populations of individual organisms — in all statistical descriptions, the continuum approximation is of great convenience and very powerful utility. The methods of combinatorial analysis or "finite mathematics" however, are now beginning to be of a more direct and *sui generis* importance.

In order to illustrate the dichotomy of such a point of view, we might, anticipating our further discussion, take up the problem of the operation of the nervous system and, more specifically, that of the brain. The analogy between some of the functions of the brain and the operations of electronic computing machines has been discussed right from the beginning of the construction of these machines. The operation of a computer is principally, and at the present time almost exclusively, serial, representable by a one-dimensional sequence of discrete operations that are taken — to be sure with very great speed — only one at a time. The course of a computation consists of many individual steps, each taken in just one of a (in practice rather small) number of alternatives. The machine operates with a number of components, each assuming a finite (in a "flip-flop," only two) sharply distinguished number of states. It is essentially this property that distinguishes the digital computer from analogue machines, which operate on a more continuous input with outputs also being continuum-like variables. These, of course, at a final stage can be made discrete, to give "yes" or "no" answers. At the present time it is not completely clear whether a digital or an analogue model is to be preferred as the main mode of description of the workings of the human brain. Undoubtedly, features of both setups are present. The operation of hormonal factors clearly presents features of the analogue or continuous schemata. In the logical part of "decisions" the discrete aspect and the sequential nature seem to be the more prominent, although even that is subject to caveats. Some "decisions" or "conclusions" may be taken, as it were, by a majority of the vote of influences or via thresholds on continuous variables. I want to stress another feature: the machines that operate temporally in sequence seem to be constructed by a discrete or a combinatorial pattern; those that operate simultaneously or in parallel on very many channels (e.g., any physical continuum, many organs in a living system) seem to be working largely in continuous or analogue or, more properly, analogy patterns. So these electronic computers and perhaps some instructions giving schemata in the cell, i.e., the DNA chain as a code, together with the transfer molecules, etc, suggest description by discrete mathematical method such as logical and combinatorial mathematics.

The operation of the brain as a whole and the working of much of the cell components and organelles would have to be described by methods similar to those used in hydromechanics, electromagnetism, plasma physics, etc.

It is really only in the *spirit* of the two classes of approach that a separation is possible, if it exists at all. Clearly, electrical phenomena are responsible for conduction of the signals, both in the computers and between the neurons. Electromagnetic waves may play a role in the actual working of the brain beyond that of being merely outward signals of the activity in the networks. Let us remember the duality of outlook in physical theories, the point of view where particles are more combinatorial or discretely postulated than in field theories, where particles would correspond to singularities or to extreme intensities of a continuous substratum or, more abstractly still, to probability waves in quantum theory.

In mathematics the fundamental ideas of elements, sets, classes of sets, etc., that is, iteration of the process of going to logical variables of higher type, and "type theory" are sharply delineated and have no counterpart in more continuous or probabilistic, more "fuzzy" mathematical models, even though some attempts in such directions are now beginning. Certainly the idea of *hierarchies* of ways of storing biological orders and, indeed, hierarchies in the very organization of the structure of a living cell, has definite substance. So, for example, the organization of memory in the nervous system very likely involves several stages of classes of variables: "sentences," "paragraphs," "chapters," etc. The recent speculations on the possibilities of our memory being organized in a "hologram" fashion certainly have some merit. The coding of patterns in the visual brain might involve more global functions of the sets of points that constitute a picture on the retina. Thus an analogue of the coefficients or Rademacher functions in a development in a Fourier series would give successively in the memory, first, information about the overall properties, and then about increasingly smaller details. Perhaps a similar decomposition into a "wavelike" sequence dealing with auditory information is also used and perhaps it is permitted to speculate that the sense of "smell" is organized by successive approximations. The process of recognition in general, be it visual or on a molecular scale (in the realm of antibodies, for example), similarly might involve *stages* of operation. We shall return to the problem of various ways of coding biological information and, later on in this chapter, we shall discuss the question of a quantitative measure of comparison, i.e., of *distance* or *difference* between such codes.

A quantitative treatment of problems of morphology could use the notion of a distance as a measure of difference between the elements of a set that constitutes the object of study. These sets, in biological sciences, are of enormous variety in both the nature of their elements and in the kind of problems that are investigated. We shall give here several examples. A fundamental one is the set of *codons* in a DNA chain. The elements are sequences of four symbols that we might call 0, 1, 2, 3. These sequences may be of very considerable length, running to some 10^9 symbols. Given a collection of such, one might want to define a measure of difference or a *distance* between any two sequences of this collection. In order to form a *metric space* this *distance* between any two elements a, b denoted by $\rho(a,b)$ has to satisfy the following properties:

1. $\rho(a,b) = 0 \equiv a = b$

2. $\rho(a,b) = \rho(b,a)$

3. $\rho(a,c) \leq \rho(a,b) + \rho(b,c)$ for any three elements a, b, c.

Our set consists of sequences of four symbols, but the mathematical problem is the same for sequences consisting of two symbols each, i.e., of 0s and 1s. The sequences may be of different length. Consider two possible kinds of steps performed on a sequence: step one is a *change* of a 0 into 1 or 1 into 0. Step two consists of *deletion* of a symbol in the sequence (after which, if the deleted symbol is in the middle of the sequence, we think of the sequence being contracted, i.e., if the original sequence was $a_0, a_1, \cdots a_k \cdots a_n \cdots$ and the deleted symbol was a_k, the new sequence will be $a_0 \cdots a_{k-1}, a_{k+1}, \cdots a_n$). Given two sequences $\{a_i\}$ and $\{b_i\}$ we can define as distance between them the smallest number of steps of our two types, performed on one or both sequences, which bring them into identical sequences. Our two types of steps are not the only or the most suitable way to express the biologically meaningful number of mutations or other changes that would appear to make the two sequences identical. In "reality," deletions and substitutions of connected *blocks* of sequences ought to be considered. In addition, our "steps" could have different *a priori* probabilities if we interpret them as due to random mutations, etc.

Another set or space of elements of interest would be that of sequences of amino acids occuring in a linear order in proteins. This space would consist of sequences of symbols, each assuming a value from 1 to 21. These again may be of variable length, even for a protein of the same kind. Again, a notion of distance between two such sequences can be introduced and a measure of similarity or rather of dissimilarity defined.

An altogether different kind of space is provided by a collection of the three-dimensional objects themselves — for example, the three-dimensional configurations of variations ("isomers") of a given protein or perhaps the collection of sets corresponding to a number of different proteins. Here the question of a distance between two three-dimensional objects can be examined by comparing the two given objects in their most favorable proximity. To explain: given a space E (for example the plane or a unit square), one considers various subsets $A, B, C,$ of this space as elements of a new space of *all* such subsets and denotes it by 2^E. If we restrict our attention to *closed* subsets of E we can define a distance ρ_H between two subsets A and B by the formula

$$\rho_H(A,B) \;=\; \max_{x \in A}\; \min_{y \in B}\; \rho(x,y) \;+\; \max_{y \in B}\; \min_{x \in A}\; \rho(x,y)$$

where the symbol $\rho(x,y)$ denotes the usual distance between any two points x,y in the plane. This metric is known as the *Hausdorff* distance.

In applications of such notions to biological situations we come upon the following difficulty in defining precisely what constitutes the elements of our space 2^E. We want to abstract from *orientation* in the space E or, in some cases, even from the *size* of objects in E that we want to compare as to their similitude or dissimilitude. Then when we want to compare and define distance between individual specimens we find that these samples are taken perhaps at different times and from different locations. As an illustration, consider the problem of defining a distance function between different triangles in a plane but abstracting from their orientation and their size. Given two such triangles we would naturally attempt to bring them to a comparable size and orient them so as to make them as close to each other as possible. This amounts, then, to defining a distance between the two *classes* of triangles $\mathfrak{A}$ and $\mathfrak{B}$ rather than the individual triangles A, B. This could be done as follows:

$$\rho^*(\mathfrak{A},\mathfrak{B}) \;=\; \max_{A \in \mathfrak{A}}\; \min_{B \in \mathfrak{B}}\; \rho_H(A,B) \;+\; \max_{B \in \mathfrak{B}}\; \min_{A \in \mathfrak{A}}\; \rho_H(A,B)$$

Here is an important point we want to stress: the objects of our study for the purpose of defining a morphological distance are *classes* of sets rather than examples, i.e., individual sets. To make, perhaps, a humorous illustration of what is involved we define a distance between the class of dogs and the class of cats as follows: given any individual dog, we find the cat most similar to it, compute the "distance" between them, and then take the maximum of that minimum for all possible dogs. We then do it symmetrically the other way around and the sum of the two numbers could serve as our "morphological distance." There are, then, two important differences between the usual definitions of distance for mathematical objects, e.g., distance between two real numbers or between two functions $f(x)$ and $g(x)$ on one hand, and between biological entities that are, as in the examples above, classes of sets. First, one difference is that in effect we compare two individual examples, not in a rigid framework, but in a most similar "orientation." Second, we really deal with classes of sets of points, that is to say, with "properties," rather than with individual examples of the sets. Thus, seemingly paradoxically, the degree of abstraction in the definition of the morphological distance between the "real" objects of natural sciences is very high — the logical variables that are used are of a high type, the *space* of various *classes* of sets. The sets themselves may be two- or three-dimensional objects, or else already quite abstract sequences of codons.

The examples above concerned two types of distances: one that is more proper for distinguishing between visual or three-dimensional pictures of three-dimensional objects, the other for a *combinatorial* measure of difference between sequences of symbols, coding abstractly a number of properties of such objects as the distance between sequences of DNA chains or between sequences of amino acids, etc.

We may mention a third type of distance function, in a sense a mixed type, arrived at as follows: In the problems of "pattern recognition" (be they questions of constructing machines that react to patterns of signals on a screen or to a retina in the eyes of a living organism or that recognize antigens of foreign substances by antibodies), the mechanism of "learning" seems to be essential: given a class of objects or of signals that are not defined logically but only exemplified on a perhaps very large number of instances, the machine (or the nervous system, or the brain) is supposed to make a decision, *given a further example,* whether it belongs to the class exemplified before or does not belong to it. One way of dealing with this problem could be as follows: the finite number of examples exhibited in the teaching or in the experience of the organism is coded, say, into a sequence of symbols — which might as well be assumed to consist of two values only — a sequence of 0s and 1s. This finite set of points in the space of all such sequences is stored in the memory and perhaps further coded by a smaller number of parameters. Now, given a further new point, the machine or the brain "computes" the distances of this point from the given set and if it is sufficiently small we might assign it to the given set, i.e. "property." It is perhaps clearer to imagine two given classes of examples illustrating two different notions or properties, N and M; then, given a new single example, i.e. a point in our space of symbols, one has to decide whether it is closer to the set N or closer to the set M.

As we hinted before, if the class of examples is presented visually, i.e. as a set of pictures, one might try to code up each of them as a sequence of numbers as follows: instead of remembering separately each individual point of the square on which the picture is presented, one regards the large screen as divided successively into, let us say, four subsquares, each one of these into four smaller ones and so on, continuing for a number of stages. The occupants of the large squares, that is to say, symbols 0 or 1, in each of the first four positions of a linear "code" for the two-dimensional picture will be used as follows: we write 1 if a square contains a point of the picture, 0 otherwise. The first four symbols will thus tell us which ones of the four large squares have points of the given set in them. The further 0s and 1s will refer successively to the squares of the subdivisions. We could, of course, attach more *weight* to the initial coordinates dealing with the overall global properties of the set, e.g., we could attach a weight of 1 to the first four weights, then 1/32 to the following sixteen smaller squares, etc. Each single example will thus be expressible by a single sequence of 0s and 1s, that is to say, by a real number in a binary expansion. A *class* of examples will thus constitute a *set* of numbers. Given a further new *point*, one can compute its distance to this set of numbers. This procedure amounts in a way to developing a characteristic function of a *set* presented visually on the screen into the analogue of the Fourier series or of a Rademacher series and to considering the set of coefficients as a number (in our illustration our coefficients were chosen to be only 0s and 1s for the sake of simplicity). A *class* of objects or a *property* is thus a set of numbers. Distance between "properties" is in this way of a mixed nature, as it were — a combination of the two ideas for distances discussed above.

One could define the degree of analogy or of similarity between two sequences of symbols somewhat as follows: we compute the distance between the two sequences and then say that their length *divided* by the distance forms a measure of similarity. In other words, the longer the two sequences and the smaller the distance between them the more analogous or similar to each other they would appear. Other more elaborate points of view are possible: we could consider again the sequences of 0's and 1's as pictures iterating our procedure; we could adopt the method of "Fourier decomposition" for these and recode our sequence again by a procedure of cutting up the whole length of a sequence into two equal parts, subdividing each into two subparts, etc.

Still another notion of distance could be of interest in problems of genetics or, more specifically, in a study of genealogies. In the simplest case we consider a population of N individuals that mate at random, and each pair produces for the next generation, assumed for simplicity to be in the next instant of time, exactly two individuals. The process is repeated indefinitely; one might assume that it has gone on for an indefinite past, the population being, with our assumption, constant. The problem is to assign a pair of individuals in a given generation a degree of their relationship perhaps imagined to be a real number satisfying the properties of a distance by considering the two sets of all ancestors (going back to minus infinity in time). One possible definition is: given two individuals x, y consider the sets of all their ancestors x and y, respectively, and define their genealogical distance (x, y) by the convention $\rho(x, y) = X \Delta Y$. Our notation is such that Z denotes the cardinality of the set Z, that is to say, the number of elements in Z, and Δ denotes the symmetric difference between two sets, that is, the set of all elements in X that do not belong to Y, together with the set of all elements in Y that do not belong to X. It can be shown that with *probability* 1, this set is finite for any pair of elements in our population, and the expected value of this distance for an arbitrary pair of elements from the same generation is equal asymptotically to $2N$.

The assumptions made about the population processes above can be considerably weakened with the result still holding. So, for example, it is sufficient to assume that the number of the offspring of a pair is two *only on the average*, etc. This definition of genealogical distance between individuals or the distance in consanguinity is only one of the possible notions for a quantitative measure of genetic relationship. Some other distances have been proposed, more suitable perhaps for measuring the degree of overlapping of the sets of genes present in two different individuals. The point that we would like to stress here is that the distance is defined in a stochastic fashion; that is to say, it exists "almost always." In order to make the "almost always" precise, one needs to have a *measure* in the space of all possible outcomes of a branching process.

Our example suggests a different approach to the problem of evolving systems. Instead of considering a *metric*, i.e., the notion of a distance between two elements x,y, one might imagine our set or space of objects to consist of outcomes of various processes taking place in time and having the character of evolution. One of the mechanisms in the process might be a random one, an analogue of random mutations, for example. Instead of considering for any pair of elements $x, y, \rho(x,y)$ with the properties of a distance, we could postulate a different notion $\sigma(x,y)$, denoting the probability of transition from x to y. This probability should satisfy certain obvious postulates, for example,

1. $0 \leq \sigma(x,y) \leq 1$
2. $\sigma(x,z) \geq \sigma(x,y) \cdot \sigma(y,z)$

for any three elements x, y, z. Indeed the probability of the change from x to z should be at least equal to the probability of having gone from x to y and then from y to z.

We now will try to give a very brief account of the application of the idea of distances in the attempts to reconstruct evolutionary trees and will formulate a rather general principle for the *a posteriori* probabilities.

A distance function can be introduced in the set of cytochrome C molecules in various species. The original work in the analysis of the linear sequence of amino acids in this protein, with the study of differences in it taken from creatures like monkeys, horses, turtles, fishes, etc, was made by Margoliash with the aim of reconstructing the evolutionary tree, leading through mutations to the diverse forms now extant. The results of Margoliash and Fitch further elaborate this work and lead to possible evolutionary trees that seem quite satisfactory to zoologists. We shall not go here into the definitions of distances proposed by various authors in this connection but merely note that some of them are quite similar to the distance between sequences of codons referred to above in this chapter (defined by a combinatorial recipe involving the minimum number of certain elementary steps that will make the two sequences identical, the steps involving both substitutions and deletions). What is crucial is the idea of introducing new, perhaps at present *nonexisting elements,* with postulated cytochromes of intermediary compositions, so to say, in order to serve as vertices of the evolutionary tree terminating in the presently observed sequences. This can be done in various ways. The problem is to find the most likely such histories. A general guiding principle can be formulated as follows:

In a metric space E a number of points are given with known distances between them. We look for a number of additional points in our metric space such that a binary tree of connections can be found terminating in the given points and such that the sum of distances between the points paired is a minimum among all possible such systems. Preferably there should be one or only a few "initial points" each associated with, at most, two others. To illustrate the nature of the problem we assume a number of points given in the space that is the Euclidean plane on our figure. The given points are circled. The points we add to presumably minimize the total sum of distances among all possible trees connecting the given points are represented by crosses. There is one "initial, ancestral point." (One could of course postulate the minimum of the square root of the sum of squares of the distances, etc.)

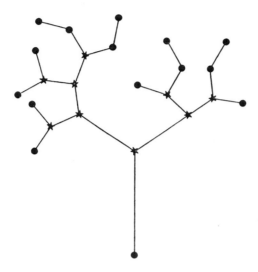

In reality, of course, our points do not lie in Euclidean space but are elements in a metric space of codons. The connections are merely expressed in terms of distances but the space consists of all possible "sensible sequences of codons." Our principle of stating that the most *a posteriori* likely evolutionary history involved entities that minimize some of the distances corresponds to a statement of least improbability. More properly speaking, since in practice the probabilities of each set of points giving our tree are very small anyhow, the solution, of course, need not be unique. There may be several sets of points giving approximately the same or even perhaps exactly the same sum of distances. A point should be stressed here. In this version our principle admits that the probabilities of a transition vary inversely with the distance between the points. The linear assumption of a *sum* of distances that has to be minimized reflects the maximum of the probabilities only asymptotically. What is most important to realize is that one makes the assumption of the mutations being not only random but also independent and neutral in this application. A more realistic but mathematically more complicated formulation of the principle would be one stated not in a metric space with distances $\rho(x,y)$ but rather in our probability space with a function $\sigma(x,y)$ alluded to before. In such a space the assumption of independence of the σs could be weakened and, to some extent in a given example, could take into account the effects of selection.

Our *principle of minimum improbability* could have wider applications beyond those concerning evolutionary trees of development of species. A simpler example would be that of a physico-chemical system of a number of substances that by random accretions, reactions, or combinations, form new aggregates leading to a variety of geometrical shapes etc.

A barest discussion of the type of nonlinear *mathematics* can be given in one or two illustrations or examples.

One of the earliest and, at the same time, most typical problems was formulated and studied by Volterra. He considered two species of organisms (fishes in his first case) that compete with each other for food and at the same time one may devour the other. The mathematical problem concerns the behavior, i.e. the changes of the population, as functions of time. We let x be the prey and y the predator. If $x(t)$ and $y(t)$ denote the numbers of the population at time t, one will have equations of the type

$$\frac{dx}{dt} = x(\epsilon_1 - \gamma_1 y)$$

$$\frac{dy}{dt} = -y(\epsilon_2 - \gamma_2 x) \qquad \text{where } \epsilon_1, \epsilon_2, \gamma_1, \gamma_2 \text{ are constants}$$

determining the process. These are ordinary nonlinear differential equations of second order (quadratic) in the unknowns. If one had more than two, say n, interacting types of organisms, one would be led to a system of n ordinary differential equations containing products of the unknown functions. Equations of this type are familiar in the theory of chemical reactions. The so-called law of mass action involves such expressions, where the nonlinearity arises from the fact that the rate of reaction (in our case, of the growth of the population) is proportional to the numbers of *both* the reacting organisms. Quite obviously, similar nonlinear mathematical problems will arise in the study of the growth of a population that increases by sexual multiplication. The number of offspring of a given type in the new generation will be proportional to the numbers of individuals in the preceding generation — both males and females — of prescribed types.

To give a purely schematic or symbolic example, we may imagine that the population consists of a large number N of individuals (both male and female) consisting of, e.g., *three* different types. The *fraction* of the population of type 1 is denoted by x; of type 2 by y; of type 3 by z. Assuming that the mating proceeds at random and that we have rules for appearance of types 1, 2, and 3 as functions of the types of their parents, we would get equations of the following form. Suppose (quite arbitrarily) that the mating or the combination of type 1 with 1 produces offspring of type 1. Type 1 and 2 produces type 2; type 1 and 3 produces type 3 again. The mating of type 2 and 2, however, produces the first type. Mating of type 2 and 3 produces type 2 and finally, mating of type 3 and 3 produces types 2. This arbitrarily chosen and rather strange rule would lead to the equations

$$x_{n+1} = x_n^2 + y_n^2$$

$$y_{n+1} = 2x_n y_n + z_n^2$$

$$z_{n+1} = 2x_n z_n + 2y_n z_n$$

where x_{n+1}, y_{n+1}, and z_{n+1} denote respectively the fractions of the three types in the $n + 1$st generation. Purely mathematically there are of course many different rules for production of types from pairs of types of parents. For three variables the number of *essentially* different such rules is 93. For four variables, the number of combinatorially possible distinct rules is 34,337. Only very few of such production rules could make biological sense.

The first problem is to find out the asymptotic behavior of the population. In some cases the numbers x, y, and z will tend with increasing time to definite limits, expressing the fact that an equilibrium state of the population is approached. In some other cases, however, oscillations will continue indefinitely. Sometimes (i.e., depending on the adopted rule of production of the type as a function of the type of the parents) the oscillations will converge to a periodic behavior. In other cases a very irregular and "erratic" development continues indefinitely. In these problems we assume that each pair produces two offspring so that the population remains constant.

One should stress how naturally such nonlinear problems come about in biological situations and how incomplete the mathematical techniques are at present. An enormous mathematical literature deals with linear problems — studies of linear operations and linear *transformations* of spaces into themselves. (These spaces may be more general than the n-dimensional Euclidean space; a great edifice of mathematical work deals with linear functionals and transformations in infinite dimensional vector spaces like Hilbert space.) By contrast, the study of nonlinear, e.g., quadratic transformations, even in the case of the finite-dimensional Euclidean spaces, is only in its beginning. The biological processes in a single organism and the processes of growth of a single organism and of classes of organisms, i.e. populations, provide here challenging new mathematical problems. In the problem of evolution, considered even in only its most schematic and rudimentary fashion, many such questions arise immediately.

One problem concerns the estimation of the *rates* of evolution in a mathematical model with many crudely simplifying assumptions. So, for example, one could assume in the first version of such a problem that the population of unisexual organisms procreates by mitosis and the increase in a given type of individual per unit of time, or in a generation, is proportional to the relative number of "advantageous characteristics" it has over the average such number present in the population. Assuming that these "improvements" are acquired by purely random mutations whose rate is assumed, one can try to compute the average number of advantageous characteristics as it increases as a function of time. Making guesses as to the number of such necessary to produce the presently existing organisms, one can have a first guess at the length of time (i.e. the number of generations) necessary to produce these by the selection process.

The next problem, still extremely schematized and simplified, concerns a population that reproduces by mating; making again very simple assumptions about inheritance of "improvements" from both parents, one arrives at a rate of acquiring favorable characteristics that is now faster and more like a quadratic function of time rather than a linear one, as in the case of unisexual reproduction. The problem itself is now nonlinear — the growth of the population and the number of new improved genes depending on binary reactions. Simplified as they are, the problems are really quite complex combinatorially and assume knowledge of a number of constants that are of course not sufficiently known, sometimes even as to the order of magnitude. This numerical work is discussed in Chapter 10.

But it is in the problem of evolution on an at first seemingly simpler level, in the question of self-organization and the evolution of biological macromolecules, that mathematical ideas on combinatorial schemata and in analysis will be useful and may even influence the concepts themselves. A deep and intriguing study by Eigen opens fascinating vistas on the processes of growth of organized graphs, the cooperative interactions in polynucleotides, self-organization, self-reproduction, and even on the origin of the biological code itself.

We now want to discuss briefly the problem of how much of the development of an organism or a part of it — an organ — is determined by a preordained code and how much of its structure and function is due to outside interventions and the influence of the "external" world. It is not easy to make this general question mathematically precise. Very likely the instructions contained in the DNA code contain some of a logically higher type: they may refer to changes of the more concrete instructions that govern the manufacture of molecules. In other words, "the rules for change of rules" might be coded in, these to become operative in response to events outside the organism but in contact with it. It might be that this arrangement could have a counterpart in a combinatorial mathematical structure which, together with the usual number of symbols, operations, and rules of inference and together with the axioms, also has a non-formalized or meta-mathematical recipe for adding new postulates along with, and depending on, the development of the body of theorems and proofs. The stimulation for such setups is occasioned by the whole development in recent years of the "undecidability," of the independence, of certain axioms in set theory.

The amount of information contained in the genetic code cannot thus be easily estimated or bounds given to it from above if some of the code should contain inductive or recursive rules. To illustrate this in a simple example: a sequence of the first, say million, primes, written down in increasing order, might at first appear to contain an enormous amount of information necessary for remembering or transmitting it. A logical and arithmetical definition of the sequence, however, would require many fewer symbols than a mere transcription. Partly because of the desire to illustrate, with geometrical examples, questions of this sort, a number of numerical investigations were started concerning recursive rules for development of simple geometrical figures.

One can define an inductive recipe for producing a growth of a plane set consisting of squares of a large (or infinite) chess board starting with one or a few initial squares as "given" in the first generation and then defining new squares of the $(n + 1)$th generation by assuming the new squares to grow from the sides of the old ones. A "contact inhibition" rule suggests itself immediately; this can be formulated in various ways. For example, one can postulate that if two of the new squares-to-be touch on a side, neither of them is put in the next generation, etc. Using very simple rules of this sort, one obtains on a plane lattice of squares (or in a hexagonal division using triangles) curious, coral-type figures whose properties are in most cases very hard to predict.

One can introduce a rule of "death"; we erase squares that are a certain fixed, arbitrarily chosen in advance, number of generations old — e.g., we decide to erase old squares that are five generations old. With this convention the growing figures will split into separate parts, sometimes resembling their parental shapes, and they would seemingly move about in the plane. What is of interest is merely the very great variety and essential unpredictability of the process, which has been defined by a very small number of rules.

For added amusement one can define contests or fights between two initially started and growing configurations. The ensuing problems of game theory are of forbidding mathematical difficulty. A short account of such experiments performed on electronic computing machines can be found in Chapters 9 and 10. A particularly elegant and ingenious game of this sort has been invented by J.H. Conway in England. He called his set of rules for producing the growing plane figures a "game of life." Some questions concerning the properties of the processes of that sort are connected with problems of decidability in axiomatic theories.

Clearly one can define analogous processes of "recursive growth" in three or more dimensions. It was very surprising to the author to observe the enormous variety of different shapes appearing from starting configurations of just a few cells, depending on their original arrangement.

At this stage of purely symbolic and extremely simplified inductive operation, there cannot be any direct comparison with the still infinitely richer class of living shapes. Some of the effects of "contact inhibition" seem to dictate growth properties of a collection of shapes and presumably their differentiation in functioning. These rules can apparently be changed by the action of certain substances.

The experiments of Puck and his associates seem to show that certain substances indeed act on cells in a "rules to produce the change of rules" manner. The experiments of Edelman on the mechanisms of immunology seem also to indicate a hierarchical structure of recognition and reaction.

We close with a mere listing of some of the areas of biological research where mathematical ideas already have or may prove useful. In statistical analysis, perhaps the oldest field of applications of quantitative work in natural sciences, the continuing importance is obvious. A field in which much useful modeling has been accomplished is that of biological systems such as blood circulations, modeling of specific organs, e.g., cochlear mechanics (of the inner ear), production of blood cells of various types, including some of the features of the immunological system. This can include systems of organisms — problems like those of slime mold migrations, etc. Clearly, with more experimental data available in the future and with the advent of more factual knowledge, modeling of the central nervous system will become more mathematically interesting and perhaps even more realistic.

General theories of systems, some fragments of which were alluded to or briefly suggested in our text above, would deal with problems of self-organization in general, including theories of communication between cells, between organs, and between separate organisms. A simulation of biochemical reactions starting with the modeling of protein interactions and going on to the reconstitution of evolutionary histories will undoubtedly attract increasing mathematical work.

At the present time it is in the field of medical applications that the computer has been found most directly useful. Keeping medical records for individual or statistical analysis, interpreting electrocardiograms and electroencephalograms with computer-aided medical diagnosis — these have already been of value. The computer representation of chromosomes with possible defects to be automatically noted is part of computer pictorial analysis of photographic biological data in general.

The specific problems in all such investigations will, in turn, suggest new mathematical ideas and techniques. The patterns of the living world will generate new abstractions.

CHAPTER 13

FURTHER APPLICATIONS OF MATHEMATICS IN THE NATURAL SCIENCES

IN THIS CHAPTER I would like to give some additional examples of sophisticated mathematics used by physicists and biologists today. There are situations where classical mathematics, the solution of differential equations, which describe physical phenomena, are not obtainable in a closed form by the methods of standard analysis. A lot of numerical work of heuristic value was done in Los Alamos during and after World War II, and this is perhaps a novel feature of present day applications of mathematics, but the physicists there also used some very abstract mathematics. For example, some results of topology were of use in the design of accelerators.

At Los Alamos, just after the war, the fixed point theorems concerning transformations of a ring into itself were useful to obtain ideas of the behavior of orbits in a circular accelerator of particles. Some modern accelerators consist of a ring in which particles, such as protons or even electrons, go around a million or many millions of times in a very short period, inasmuch as their velocity is close to the velocity of light. These orbits in the torus have to satisfy certain properties — they should not get too close to the walls. It turns out that topology theorems concerning such transformations of a torus into itself have direct applications, probably something that a pure mathematician would not suspect *a priori*.

There was at Los Alamos from the beginning of the project a group devoted to the study of the effects of radiation, specifically the study of the health hazards of radiation. This group started with very specific problems concerning the effects of gamma rays, neutrons, and charged particles, which are very dangerous to the health of people working near accelerators with radioactive materials. People working with such have to wear film-badges that record the amount of radiation they are being exposed to. This group contained a number of biologists and medical people. In the beginning, it was a rather small group, but when the war ended it became a whole division where one studied not only the immediate practical problems of the hazards and dangers of radiation and its effect on tissues, but more fundamental biology in general. They studied problems of genetics, and in order to put the knowledge of these problems on a more general basis, they studied properties of cells and their reactions to external influences, in particular those of radiation.

Now to pursue this somewhat autobiographical vein, I should mention that in talking to some of the members of that division I became more and more interested in the problems of biology in general. This was in the late forties and early fifties, before the great discovery of Watson and Crick on the mechanism of replication and the role of the code that describes the development of living organisms. This discovery, by the way, besides its enormous direct import on biology in general, also had a psychological effect on biology. It revolutionized the aspects of molecular biology and even influenced more generally a whole field of natural sciences.

For a long time, I was interested, in an amateurish way, in biology. When I tried to ask general questions of some colleagues, at Harvard or other places, some 45 years ago, the answer was "Ah, you cannot say this because there is an exception. What you say is sometimes true, but there are such animals, or insects, or fish that are not that way." Even slight generalizations almost seemed to be resented in those days. The discovery of Crick and Watson which showed a certain universality, a great generality of a schema of replication, greatly changed the attitudes of biologists and people in natural sciences. As a consequence of these ideas, there is now much more willingness to talk about generalities, to speculate on generalities that might be discovered, and on the theoretical variations of what we, in a special case here on earth, call life. I cannot explain here or give examples of this revolution in attitude, but there certainly has been a change in the psychology of the researchers.

Some time after World War II ended and I returned to Los Alamos, I held discussions with biologists in the so-called H Division (H for health), and with some physicists and a mathematician or two. We organized a little seminar for laymen like ourselves with some professional biologists who were willing to tell us about the fundamental properties of the units of life and teach us about the morphology, behavior, and role of cells. I certainly did not know enough organic chemistry to discuss this thoroughly, but really was interested in the general conceptual problems, problems of the schemata of the organization of the cell and its functioning. We noticed that indeed some simple mathematics could be useful, not merely as a service function to help calculate by differential equations phenomena that were understood in principle or physically, but perhaps even in formulating the schema of the organization and functioning of the cell itself. To this day these schemata are not completely known, and fundamental questions remain unanswered.

Of course, one knows that mathematics, that is to say statistics, was always useful in biology. Even in the fundamentals of Mendelian theories one needs a little probability theory. There are the simple and nice observations of G. H. Hardy, a pure mathematician, a great number theorist. There are studies using differential equations to describe the flow of blood in veins and arteries; the idea of diffusion. In other words, classical mathematics had been used for a very long time. But it seemed to some of us, to myself certainly and I still feel it very strongly, that the time had come for a deeper involvment of mathematical work on a conceptual basis, on the basis of the formulations themselves, to the problems of biology.

I would like to talk now about the possibility of using abstract, more general mathematical ideas in the schemata of molecular biology, and in problems of genetics and evolution. I will also make a few remarks about the problems of organization of the nervous system and of the functioning of the brain itself. All this, of course, is quite different from the application of mathematics to theoretical physics, because much less is known about the facts of biology in the areas that I have mentioned; although during the last couple of decades exploration is proceeding in a most impressive way.

As I stated above, the discoveries of Crick and Watson involve the fact that a living organism is programmed, coded by a linear sequence of letters. Biologists denote it by *A, C, T, G.* This linear sequence and code is contained in the nucleus of a cell, in the chromosomes, and is very long. Now, instead of letters, we can again use symbols such as 0, 1, 2, 3, or for purposes of illustration that will not change the gist of the matter, just the two symbols 0 and 1, and consider the code consisting of a long sequence of binaries. In every organism, even in bacteria, this code in the nuclear tape is quite long. For a mammal, such as a human being, it is apparently some billion units long.

To impress mathematicians, who in general avoid complexity as such, we should point out that even a single bacterium has up to ten thousand different enzymes and proteins, each of which is coded by a lengthy expression of these symbols. The code involves not only the chemical substances, the hardware so to say, but also the coding of functional uses; not just sets of molecules, but something that corresponds to a higher logical order of instructions for motions for organization and such; this still is not understood very well.

Thousands of proteins are now decoded in the sense that one knows the linear sequence that defines them. What happens with the code is this: molecules go along the code or part of the code in a chromosome, to make a simile, in the fashion of a zipper.

These molecules read short segments of it, then travel from the tape to little spheres called ribosomes, which are in a sense factories. This transcript of a section of tape, when deposited in the ribosome, will start manufacturing the particular chemical molecule read from the code and deposit it. Substances like hemoglobin and myoglobin and many of the other important molecules are manufactured in this manner. Of course, this description of the process is much simplified.

I should point out here that there is one special substance, a protein called cytochrome C, that is present in mammals, fishes, insects, and in seemingly all living organisms. Although not identical in all organisms, large segments of the molecule are chemically homologous. Cytochrome C is important for the phenomenon of electrical conduction inside the living organism and is absolutely essential for life. The code for it is some hundred to hundred fifty units long. (When I speak of units I mean the DNA units, not just the 0's and 1's. In symbols reduced to 0 and 1, one needs a couple dozen times more units.) I also neglected to state that the code is formed by triplets of these letters, which can be thought of as forming "words." Combinations of these words, of which there are some twenty, form "sentences." These sentences then are the definitions of the proteins and enzymes and other substances that control the life processes. The universal occurrence of this protein in living organisms suggested to the biologist Margoliash the idea of trying to devise the sequence of evolution from hundreds of millions of years ago to the present, by looking at the variations in this protein. Margoliash assumed that the more complex or varied definitions of cytochrome C came later as a result of mutations. If there was a primitive original form, then one could conceivably follow its succession of changes until one arrived at its present variety. Margoliash's idea was to discuss the evolutionary tree for this molecule and to assume that it mirrored in a way the evolution of living organisms. In all, he analyzed the protein code for approximately 20 or 30 different species. Assuming the simplest one corresponded to a proto-organism, and noting the changes and differences between them, he ordered them in a tree that described successive mutations due, in all likelihood, to selective pressures. These led to different cytochromes in the more complex organisms, and the idea was to find an order in the tree based on the following assumption: small changes are more probable than large changes. Now what is small and what is large depends in our language on the distance or metric in the space of codes for this special cytochrome. One also had to assume that nonextant organisms were points in the tree. These junction points or vertices are conjectural. It is done in such a way that the sum total of the distances through the whole tree of the codes is as small as possible, that is, the sum of the distances in the sense of the metric in the space of codons. Therefore, mathematically the problem was: given a number of points in the space of codes how to construct a *binary* tree, a tree splitting into two branches at each node, corresponding to single mutations, in such a way that the sum of the distances of the edges between vertices, including of course those that do not exist any more, be as small as possible. At the end of the tree we have the existing species examined by Margoliash and his collaborators, about thirty points (theoretically one could have many more.)

It occurred to me at once that given an idea of distance (one which I will define below), the problem becomes a variation or a generalization of an old geometric problem of Steiner's, the nineteenth century geometer: what are the smallest, most economical connections in the form of a binary tree linking points lying in the Euclidean plane (with the distance being the Euclidean, normal distance used for measuring separation of points in two dimensions)? This problem is not yet completely solved even in this special form.

In our situation, we don't have a *plane*, but a general space, a space of codons, of sequences of 0 and 1, between which a special distance is defined. One wants to find new points and to connect all of them in such a way that the total sum of all the distances is as small as possible. This means that the total of the mutations was a least improbable one among the possible series of such.

Now, what kind of distance should we consider between the codes for DNA that define the organic substances? To simplify again, I will assume that the symbols are only 0 and 1, not the four letters used by biologists, although it makes no difference at all mathematically. One distance between two such sequences could be the well-known sum of the absolute value of the differences between the two sequences

$$\alpha = [\alpha_1, \alpha_2, \ldots, \alpha_N] \text{ and } \beta = [\beta_1, \beta_2, \ldots, \beta_N]$$

namely

$$\rho(\alpha, \beta) = \sum_{i=1}^{N} |\alpha_i - \beta_i|$$

used in mathematics and in physical applications. Sometimes, this distance is called a Hamming distance; really, it is a special case of the Minkowski distance, which is a more general one than the Euclidean distance between points in a plane (or for that matter, in three-dimensional space or n-dimensional space.) But this distance is not suitable for examining similarity (or dissimilarity), analogy, or the measure of "likeness" between biologically meaningful structures. The reason is this: having a code, if you imagine it as a physical object, then its position is not fixed, rigid, as is the case for physical objects lying in space. A tape can be twisted, shifted, curved, and the idea of similarity or closeness of form between two such tapes of codes would be different. For example, take the sequences

$$\alpha = 010101\ldots\ldots1 \text{ and}$$
$$\beta = 101010\ldots\ldots0$$

the two sequences are really very similar. The normal "Hamming" distance between them would be very large, equal to N if there are N symbols in each, because they differ in every *place* or index, the index being rigidly fixed for each. However, by a *shift* and one or two *erasures* one can make them identical. The obvious thought then is to modify the idea of distance to incorporate such possibilities of allowing small changes of a different type in order to compare two given sequences of 0 and 1.

A new definition to satisfy this feeling could be, given two sequences,

$$\alpha = \alpha_1, \alpha_2\ldots\ldots\alpha_N$$
$$\beta = \beta_1, \beta_2\ldots\ldots\beta_N ,$$

operations of two types are allowed. Type I is to change 0 into 1 or vice versa, and each such step costs one unit. Type II is to erase a letter and then contract the rest so as to annul the symbol that we have gotten rid of, a sort of omission of a symbol (by mutation so to speak). This operation is allowed on the first or second sequence or on both. We have then steps of two types, and we can call a distance between the two sequences the minimum number of steps that will result in the two sequences being identical. This minimum number, it turns out, satisfies the property one wants to have for a distance or a metric. First of all, the distance has to be symmetric. When it is 0, it means the sequences are identical. Secondly, we have to have a triangle inequality, meaning that from a sequence A to a sequence C the distance between these two sequences is not less than any intermediate path going from A to B and B to C. The sum of these two is equal to or greater than the direct distance between A and C. It turns out that our definition satisfies this property. Of course the original attempts made by biologists did not necessarily satisfy this last postulate of the so-called triangle inequality.

Now I want to mention some mathematical problems that are suggested by that sort of definition of distance between codes or one-dimensional sequence problems that might amuse a pure mathematician.

The first problem is an elementary one. Suppose I have two sequences given at random of 0 and 1 each of length n. What is the average distance, in our new sense, between them? It is obvious that the classical, ordinary distance, the sum of absolute values of differences for two randomly chosen sequences of length n will be $n/2$, because on the average, the difference will be in half the places. In our distance, the number, the average will be different; it is clear that it will be less than $n/2$.

At first I was trying to prove a simpler statement that this average is a linear function of n. I did not find a proof of it, but a young mathematician, Joel Spencer, professor at Stony Brook on Long Island, proved that indeed for large n the average distance tends to behave linearly with n and in the limit it is a linear function of n with a coefficient less than $\frac{1}{2}$. He does not know its exact value but has upper and lower limits for it. This theorem proves, as mathematicians would say, a weak law of large numbers for such random sequences.

For infinite sequences we may now proceed as follows: One way to define the distance between two sequences is to write them down, compare the first n segments of both, compute the distance between them and divide this by n. By virtue of the result of Spencer, this should be on the average a linear function of n. Then we go to the limit for n tending to infinity. In order to prove that it exists for almost all sequences, one has to identify any two sequences such that the distance between them, defined in the way I just indicated, tends to zero. Therefore, we will obtain a space of classes of sequences of 0 and 1, a rather interesting new type of space. I should mention here that Peter Sellers, a professor at the Rockefeller University in New York, proved nice theorems about distances between codes of this type.

What I have just said is extremely schematized and simplified. The operations we allow of changing letters and erasures are very specialized; in reality, we might make definitions more general. The changes of substitutions of letters should perhaps not be weighed equally. Some mutations are easier for certain symbols than for others. Also the mutations may involve a whole small block of letters together, not just a single one, and the cost of erasure of a block need not be equal to the number of letters erased, but could perhaps be smaller in some cases. Suppose we erase five letters, it should not cost five units, it could cost less, depending on which letters and which combinations they are. Let me also add that one can talk about additions, not only erasures of symbols, or intercalations in the middle. Mathematically, it would not make much difference. For the sake of simplicity it is easier and sufficient to talk about diminutions or erasures of symbols. To find the exact cost of erasure of a block would depend on the nature of the symbols, the biological facts, and chemical or physical properties. This has yet to be studied experimentally.

The trees of evolution that Margoliash and Fitch constructed seemed to satisfy zoologists. The zoologists have their own way to guess or estimate the possible succession of evolutionary changes among species. Recently, attempts have been made to use other proteins, not just cytochrome C, to try to define distances between corresponding sequences and use those proteins to construct a tentative or conjectural tree of evolution. There is some literature on those problems and I myself and collaborators have written some mathematical papers on such problems.

So here is just one illustration of a problem arising from the world of natural sciences that suggests mathematical schemata slightly different from some that have been considered in pure mathematics or in mathematical physics.

So far we have discussed briefly a space of codes, one-dimensional sequences of symbols. More general and more difficult analogous problems are suggested by similar considerations concerning objects in two or three dimensions, the real physical objects. A protein that is manufactured by the parts of the cell receiving the code is not a straight linear sequence but a three-dimensional object looking like a very contorted pipe. For several of these proteins, it has been possible to map the three-dimensional positions of the atoms forming them. This was much harder to obtain than the linear sequence of the DNA defining it.

Techniques of crystallography and X-ray diffraction have enabled the biologists to discover in some cases the position of the atoms forming the complicated large molecules corresponding to certain proteins. Some hemoglobins and myoglobins, for instance, were learned in this manner. Mathematically, the technique depends on the inversion of Fourier transforms, a very laborious analysis where computers are absolutely necessary to enable one to discover what the distribution of atoms is in some of these large assemblies. As I said before, the linear code is known for many hundreds or thousands of proteins. New techniques of cutting up the linear sequence are permitting an examination of the complicated sequence of symbols by various difficult chemical techniques. It is more difficult to study the three-dimensional structures, but a number of them are known now. A few years ago when I looked at an atlas of proteins there were less than 100 such spatial shapes.

Now one might try to discuss the possibility of defining the degree of similarity or dissimilarity or analogy between *spatial* objects of this sort. How does one measure quantitatively, or at least somewhat quantitatively, the degree of similarity between the shapes of organisms? In general, these are three-dimensional objects, and in mathematics, in pure mathematics, people have discussed for some time the ways to define distance between two sets. Suppose these sets are in the Euclidean plane or Euclidean three-dimensional space.

The mathematician Hausdorff defined a measure of distance; he made a metric space out of sets, say closed sets in a given metric space. I shall now briefly describe the Hausdorff distance between such sets A and B. We take a point on A, call it x and look for the closest y in B. We take the distance between these two points x and y. Distance is defined in the general metric space in which these sets are located. Then this y is the closest to x. We take the minimum for all y with respect to x. Then one considers the value of this minimum with respect to the original choice of the point x and we take the maximum of that. If the sets are closed and bounded this maximum exists. So the maximum of this minimum we define as distance. I should add that in order to make the distance symmetric we turn the expressions around, we start at a point belonging to B, look for the closest in A and take the maximum with respect to the initial point in B. The sum of these two we consider as the distance between the two sets A and B. It satisfies the common sense requirements for a distance between sets. All this assumes, however, that the two sets are fixed, rigid, positioned in a given metric space. Again, when it comes to objects that are organisms, i.e., sets that are not really fixed in space but are pliable, soft, malleable, deformable, and can possibly be twisted, this procedure is not the right one.

One should actually look not at a fixed set A, but at all the sets like A, be they translated, rotated, subjected to small deformation, and so forth. That is to say, we have a class $\mathcal{A}$ of sets that are "like" A. Then having another set B, we again do the same and have a class $\mathcal{B}$ of sets like B. What we want to define is a distance between two of these classes. In order to make it a bit extreme for illustrative purposes, suppose that A is a dog, B is a cat. I then consider something that I would like to define as the distance between a cat and a dog as species. It amounts to taking all possible sets like A, all possible sets like B, and attempting to define the distance between these two classes. This will be the Hausdorff distance. It turns out that this iteration of the Hausdorff distance leads to a rather reasonable idea, it seems to me, of distance and a metric space between closed classes of closed sets of a given type. I shall not go into details of how the classes are precisely defined but merely say that mathematically it amounts to allowing certain operations that deform or transform the given sets A and B. They are not groups of transformations, that would be too much, rather they are kernels or neighborhoods of unity of certain groups of transformations or distortions, and now we can define the distance as follows

$$\rho(\mathcal{A}, \mathcal{B}) = \max_{B \in \mathcal{B}} \min_{A \in \mathcal{A}} \rho_H(A, B) + \max_{A \in \mathcal{A}} \min_{B \in \mathcal{B}} \rho_H(A, B) ,$$

where ρ_H denotes the Hausdorff distance in the space E where the sets are located, and

$$\rho_H(A, B) = \max_{y \in B} \min_{x \in A} \rho_E(x, y) + \max_{x \in A} \min_{y \in B} \rho_E(x, y) .$$

This gives another example of a mathematical technique, a mathematical point of view and a new type of construction suggested by a "real" problem of natural sciences other than physics or astronomy.

A similar technique or similar problem is found in an entirely different connection. The problem of recognition, let us say recognition by a machine, by a computer, of signs or objects on a screen or in three-dimensional space. It is of interest when we study the vast area of questions about the working of the nervous system and of the brain itself. I can only talk here about generalities. From the beginning of the construction of electronic computers one has talked about its memory, about the various programs that enabled one to execute a number of mathematical operations. By now we have a vast literature dealing with the similarities between some rudimentary properties of the nervous system and our thinking and the functioning of computers.

As everyone knows, a computer is vastly faster for all the operations that it is able to perform than the nervous system or the human brain. But it is very limited in what it can do compared to the human brain, or even to the brain of certain primitive animals, in the following way: computers that now exist work essentially in series, linearly, one step at a time. There seems to be no question that our brain, or even the brain of much simpler organisms, works differently in that the process operates in parallel, essentially simultaneously, independently on thousands or even millions of different channels. The retina of the eye has several million rods and cones connected to nerves. Behind the retina, there exist four or five layers where there is a rewiring. This scheme is not yet really understood in detail or perhaps even in a general combinatorial sense. This rewiring finally leads impulses to the brain, which contains, in the human case, an assembly of some ten billion or more elements, the neurons, and, perhaps even more impressive, hundreds of thousand times more connections between them. I remember discussing this with von Neumann years ago, and he said "It is fantastic, from each neuron there are maybe a hundred connections to other neurons and in the central part of the brain there must be several hundreds leading to other elements." By now these hundreds are known to be many thousands, and in the central part, apparently from a single element on the axon, there are perhaps a hundred thousand wires leading to other elements.

The problem of recognition, visual recognition of a pattern, is therefore to be studied in a vastly more general way than that of a linear mechanism for processing individual single impressions step by step. I should like to add that the problem of recognition does already exist on a different level. All the problems involving immunology, recognition of antigens by antibodies, involve some chemical schemata for finding spatial patterns or shapes by a method not understood to this day, but which undoubtedly involves a parallel gathering of information and a suitable code for action.

Such problems form a very fertile field for the intervention of mathematics on a general, abstract, conceptual scale. From the beginning of the construction of electronic computers there was talk of learning machines and speculation on how to imitate the workings of the brain on a very modest scale.

In this set of problems, again, it seems to me that it makes sense to talk about the problem of distances. Suppose we have a visual set of points on a screen of our visual field. I want to discuss how one such impression of a visual picture can be compared to other pictures, and how this comparison could involve again the idea of a distance between such sets of points. These sets are not rigidly, stiffly precise, but speaking mathematically, are modulo a number of transformations. It is obvious that the picture that I recognize need not have a fixed size — it can be large or small provided it is similar. If I am at a distance or close, the same object will appear in a different way as a set of points on the retina. We recognize it, however, as the same one. In addition, translation of a picture does not alter its recognition either. Small changes in shape also are neglected in the recognition of the class of the object.

For a simple example, consider the problem of recognizing a handwritten letter, *A* for example, from a handwritten letter *B*, independently of whether it is large or small, slightly shifted, turned or written by persons whose handwriting differ markedly. In order to give a mathematical setting to this problem of recognition, I would like to introduce again an idea of distance between classes of sets corresponding to the letter *A* and the letter

B, or for that matter any other letter. We tried this in Los Alamos on a computer, and it turns out that some types of distances, one of which is the Hausdorff distance discussed above, were very useful for setting up a program that enabled the computer to decide, with a probability that turned out to be very good, the distance between the two alternatives.

Let me give a brief account of this story. The work was done years ago in Los Alamos with Robert Schrandt, (in the mid-seventies, I believe.) We obtained a number of handwritten letters A, which were put in the memory of the machine. Actually, instead of writing the letter 128 or 256 times (we wanted to have that many examples in the memory), we did one or two by hand, and then performed small deformations by computer by iterating two given transformations in turn seven or eight times. These iterated transformations deformed in various ways the picture representing the letter A. We did the same with the letter B and had it also in the memory of the machine. Then the problem was given to the machine. Somebody wrote a new letter and the machine was to decide whether it was an A or a B. In order to render the machine able to do this we introduced a distance between pictures. This distance was different from the Hausdorff distance, but I will not attempt to describe it here. There are different types of distances, different types of metrics that one can introduce in the space of pictures which are suitable, and probably all independently used for recognition by the nervous system. I should not say that these distances, which we considered, are actually used by the brain, but distances somewhat like these I think may be involved in our recognition patterns, perhaps also in recognition by animals.

Now, given a new letter, the distances were computed by the machine between this new letter and all the examples of A on one hand and then all the distances between it and the letter B on the other. Whichever came out smaller in the average of these distances made the machine decide in favor of it. The program was quite successful. There was 80 or 90 percent success in recognition. This was just a first crude attempt to discuss recognition by a mathematical program, and no doubt more sophisticated and suitable distances are used by living organisms.

The above is only the simplest problem — it is recognition of a picture and not of an object. Assuming we should be able to do that for a large class of objects, the next stage is more sophisticated. We have not a single object and its variations, but rather a class of pictures. For example, we have a tree. It need not be a specific tree, like an oak, but rather the class of trees. On the other hand, we have examples of animals. The question is how to distinguish between these classes of objects? Classes for which elements are sets are not necessarily variations of each other. This is the next step in sophistication. Such a class can be called a concept. One can again iterate this procedure and imagine classes of classes, or collections of classes. Probably in the future one will be able to devise methods for such increasing abstractions, and perhaps people who now work in computer technology and theory will be able to devise ways to recognize and discern more than the mere arithmetical or Boolean operations computers are so good at now. We shall be able to define analogues, at first still modest for years to come, of some of our thinking processes and ideas obtained by abstractions, as a better way to study impressions. It seems to me that the idea of distance between sets, classes of sets, and so on, will play a role in this program. There are many mathematical problems relevant to this idea.

The thinking process involves the examination of analogies obtained from external impressions, visual and auditory, and from the other senses. I am reminded of Norbert Wiener's and von Neumann's debate of years ago. Von Neumann saw the analogies with computing machines and Wiener was for hormonal, continuous fluid type relations as the mainstay of our thinking process. No doubt future development of computers will involve all kinds of distances between sets of impressions and later an analysis of functionals of what the external world produces as our stimuli will form more abstractly what we call concepts or ideas.

To finish this point, I would like to stress again that mathematics not only applies itself, as it is and as it develops, to other sciences, but perhaps has a direct biological role for the human race. Mathematics, and perhaps other sciences like physics, have the mission to prepare or improve the human brain, be it the brain of an individual or the collective brain of mankind, for developments yet to come. Just as animals play when they are young in preparation for situations arising later in their lives, it may be that mathematics to a large extent is a collection of games. In this light, it has the same role and may be the only way to change the individual or collective human mind to prepare it for a future that nobody can now imagine.

As biologists now discuss it, life appears *inter alia* as a sequence of chemical games. Games played between individuals, or between groups of individuals are something that is essentially of a mathematical nature. I do not mean the von Neumann-Morgenstern theory of games, but more general games in the widest sense. This has perhaps, I will repeat it again, a direct biological role.

There appeared recently a book written by the German biologist, Manfred Eigen, entitled *Das Spiel* (The Game). It describes a number of mathematical games or puzzles and discusses the games molecules and groups of molecules play with each other.

Other interesting mathematical puzzles have appeared in the last twenty years — Conway's Game of Life, for example. In this connection we have already seen that starting with a simple pattern and simple recursive rules can lead to unbelievably complicated configurations. The configurations, though defined by a rule that a grade school child can understand, defy analysis *a priori*. There are many mathematical problems beyond just amusement or diversion that involve new combinatorial patterns and constructions. One can imagine a little branch of mathematics (and call it, perhaps, "Auxology") concerned with the geometric properties of recursively growing figures.

Chapter 14

THERMONUCLEAR DEVICES

When Bethe's fundamental paper on the carbon cycle nuclear reactions appeared in 1939, few, if any, could have guessed or imagined that, within a very few years, thermonuclear reactions would become an object of most intensive theoretical studies — and that such reactions would be produced on Earth and enter decisively into the new technology of the nuclear age. The work of v. Weizsäcker on the same problem and the papers by Atkinson and Houtermans on deuterium reactions have, together with Bethe's work, drawn attention to the schemata by which energy may be produced in the stars and in our sun; but it was hard to imagine that, in such a short time, human technology would successfully dare to imitate such processes on earth. It is strange to realize now, some quarter of a century after those heroic times, how rapid was the tempo of the developments in both the theory and in the technology, developments which now in these fields proceed at a more measured pace.

As often happens in the history of science, and of physics in particular, it was an additional and rather independent development, coming almost simultaneously, which combined with the theoretical ideas on thermonuclear reactions to make them realizable and applicable. The discovery of fission and the fact that more than one neutron is liberated in this process immediately suggested the possibility of a new type of chain reaction leading to an enormous release of energy which, almost instantaneously, would appear in thermal form and produce situations where a small, terrestrial object would be instantly heated to temperatures orders of magnitude higher than any heretofore attained in a piece of matter in bulk. The story of the Manhattan District Project, of the Metallurgical Laboratory in Chicago and of Los Alamos has been told in many accounts; but it is not perhaps very well known that a small group of physicists, Bethe, Konopinski, Oppenheimer, Serber, and Teller, gathered in Berkeley during several weeks in the summer of 1942, before the Los Alamos Project was started, in order to discuss what could be done toward a design of a thermo-nuclear device. This was before the production of a nuclear fission chain reaction in Chicago; the courage of these physicists and their faith in theoretical thinking is, indeed, to be admired.

This chapter will attempt to sketch the subsequent history of the studies which were involved in the establishment of thermonuclear reactions and in the explosive release of their energy. We shall not attempt here to discuss the thermonuclear devices whose study forms the subject of the Project Sherwood, that is, the gradual release of confined thermonuclear energy and the various arrangements which have been proposed to this end.

The work at Los Alamos started in 1943 and was directed toward the design and construction of a fission bomb. The preoccupation with and the innumerable experiments on the fission process and associated phenomena — the theoretical work on the process of the chain reaction — all had some good, if only general, background of previous work. Some of the most useful that existed in published form were the great papers on nuclear physics by Bethe (and Bacher and Livingston) which appeared in the *Reviews of Modern Physics* in 1936 and 1937. They became a true compendium of knowledge in this subject and were referred to as "The Bible" for information on nuclear reactions. Alongside of all this, a vast amount of work was going on, both theoretically and experimentally, on problems of chemistry and metallurgy vital to the construction of a device which, once initiated by neutrons, would explode yielding some meaningful fraction of the total energy available in the material. The study of the process of nuclear explosions with durations much shorter than any which had been previously studied for such problems and involving the thermodynamical and hydrodynamical behavior of matter at temperatures and energy densities far outside those previously encountered on earth, preoccupied almost exclusively the scientists and engineers on the project.

The task of constructing a fission bomb was not only one of immediate urgency and the primary assignment for the laboratory, but its solution was indeed a necessary condition for any attempts to conceive, design, and plan releases of energies which could be obtained from thermonuclear reactions. Nevertheless, from the very beginning of the Los Alamos project, a relatively small group of physicists devoted most of their time to specific work on a thermonuclear bomb. "Super" was the code word for this enterprise. It was Edward Teller who directed with enthusiasm and fantastic energy the thinking and the calculational work toward this end. It was realized from the beginning that temperatures of the order of tens of millions of degrees were necessary to provide an initiation for the thermonuclear reactions. Deuterium, which could be made available in sizable quantities, was, of course, the material primarily considered. The role of tritium as a product of the $D + D$ reactions was immediately noticed; and, as a matter of fact, the possibilities of using some tritium from the outset for facilitating the progress of the thermonuclear reaction were suggested already during the Berkeley Conference, mentioned above. The problems of igniting a mass of deuterium appeared very formidable. An understanding, in greatest detail, of the details of the preceding fission explosion was necessary for the establishment of the initial conditions for the explosion of deuterium. The subsequent interaction of the effects of the fission explosion had to be foreseen and calculated.

As regards the process of the thermonuclear reaction itself, all the questions of behavior of the material as it heated and expanded — the changing time rate of the reaction; the hydrodynamics of the motion of the material; and the interaction with the radiation field, which "energy-wise" would be of perhaps equal importance to that of the thermal content of the expanding mass — had to be formulated and calculated.

Several schemata for the arrangement of the material were proposed by Teller and his collaborators, and these had to be scrutinized by extensive analytical and numerical work. It was realized very early that even to obtain a "yes" or "no" answer on the possibility of realizing a thermonuclear explosion depended crucially on being able to obtain accurate numerical descriptions of the extremely complicated processes referred to above. The schemata of the "Super" appeared feasible, even though not certain of realization in the months preceding the first nuclear explosion in New Mexico in August of 1944. To realize nowadays the magnitude of the problems involved, one should remember that, even only mathematically, the problem of the start and explosion of a mass of deuterium combined a considerable number of separate problems. Each of these was of great difficulty in itself, and they were all strongly interconnected. The "chemistry" of the reaction, i.e., the production, by fusion, of new elements not originally present, and the appearance of tritium, He_3, He_4, and other nuclei, together with the increasing density of neutrons of varied and variable energies in this "gas," influences directly the changing rate of the reaction; and it does so also by changing the values of the density and temperature. Simultaneously, the radiation field is increasingly present and influences, in its turn, the motion of the material. The work of the "Super" group on the visualization and on quantitative following of these processes constituted a veritable monument to the imagination and skill of theoreticians. We have to remember that electronic computing machines were not yet in existence in those days and daring simplifications and some, only intuitive, estimates had to be made. An account of some of this work by Teller and his collaborators was given in a series of lectures by Fermi toward the end of the war.

The work at Los Alamos on the "Super" continued after the end of the war. There were a few special meetings devoted to ascertaining the plausibility and feasibility of the arrangements and the devices as they were proposed during the war for the "Super."

Following the directive issued by President Truman to proceed with the planning and construction of an H-bomb, the work at Los Alamos, which had been generally directed toward fundamental problems of feasibility, was stepped up considerably and reoriented toward the problems of specific device design. Even before, though, certain doubts had arisen about the practicability of the schemes outlined during the war and elaborated in subsequent work. A very detailed and comprehensive calculation was planned to be performed on the newly available electronic computing machines, of the whole course of the ignition process of the thermonuclear reaction and its subsequent course. The plan was to do it as completely as possible, taking into account all the pertinent phenomena, and follow in a great number of time steps the interaction of the numerous variables. Von Neumann participated with Teller, the writer, and several other members of the Los Alamos Project in planning this gigantic calculation. Quite independently, however, the present writer, in collaboration with C.J. Everett, had undertaken the calculation of the ignition process and the following course of the thermonuclear reaction by using numerous simplifications and guesses as to the values of certain multidimensional integrals defining the distribution of neutrons and the products of reaction during the changing geometry of the mass of the active material. These calculations were performed with the aid of desk top computers worked by hand. The results of this work showed a very "weak" process. The mechanism of ignition, as considered in the existing schemata, was submarginal and led to disappointingly weak initial conditions and, after a time, to decreasing rates of reactions.

Shortly after that time, another calculation was similarly performed by Fermi and the present writer; we considered the "next" problem: Assuming that in some way an ignition of the mass of deuterium — perhaps supplemented by quantities of tritium — could be made satisfactorily, how would the subsequent reaction proceed in a volume of deuterium? Again, the calculation had to rely during its course on intuitive estimates of the properties of distributions, e.g., the neutron density, neutrons of at least two different classes in energy being products of the thermonuclear reactions; the distribution of gamma rays, i.e., the photons constituting the radiation field, etc. The result of this work was, again, quite negative in the sense of showing that, most likely, even if the initial reactions were established, as in the scheme proposed, it would not continue and go anywhere near to completion or even sizable burning of the remaining material. Shortly afterward, the calculations which were proceeding in the meantime on the newly developed electronic computing machines, and in which some of our intuitive arguments were replaced by full numerical work, were completed; and the results confirmed those of the above work.

We mention all this merely to indicate how sensitive the whole complicated arrangement constituting the "thermonuclear device" was to the interacting process determining the burning of the deuterium. In fact, the reports by Fermi and the the writer indicated that, if the cross sections for the $D - D$ reactions were between two to three times greater than they actually are, i.e., than the measured ones, the process, as proposed during the war, would indeed go. In a sense, therefore the thermonuclear device, i.e., the proposed H-bomb, was not a trivial object to construct even after having the fission bomb available!

For the wartime schemata for the "Super," the hydrodynamical disassembly proceeded faster than the buildup and maintenance of the reaction. For the thermonuclear devices to produce a meaningful amount of energy, there are properties of the assembly which play a role similar to that of a critical or supercritical mass for a fission bomb. Some new combinations of ideas had to appear before the successful thermonuclear device could be designed. A Los Alamos report by Teller and the present writer outlined a new approach. The theoretical estimates and the subsequent calculations based on this scheme were far more promising; and, as is well known, successful H-bomb designs were produced and tested quite soon after the appearance of these new ideas, thanks to the energetic and imaginative teamwork of a whole group of Los Alamos physicists. Subsequently, numerous technical improvements in sizes, weights, etc., were made by scientists at both Los Alamos and Livermore.

The devices permit enlargement by an enormous factor of the amount of energy which one can obtain from nuclear materials. Deuterium being readily manufactured, the scope of possible technological uses is vastly increased. It is true that, so far, at the time of this writing, the release of large amounts of energy from thermonuclear reactions is possible only in an explosive form — the problems of the Project Sherwood not having been as yet completely solved. We would like to emphasize a number of facts pertaining to such explosive releases.

So far, a certain fraction of the total energy from thermonuclear reactions involves fission reactions occurring at the initiation of the thermonuclear reaction. As is well known, research is proceeding toward making this fraction negligibly small. The thermonuclear reaction produces, of course, neutrons of various energies, including 14.2 MeV neutrons. These neutrons interact with the nuclei of the surrounding material and originate radioactive products. One can try to minimize the residual radioactivity of the products, taking care to select special materials for the environment of the thermonuclear device.

In the few paragraphs that follow, we shall try to outline some of the possible scientific and technological applications which are possible through the use of such releases of thermonuclear energy. The phenomenological characteristics of thermonuclear explosion involve a great range and impressive values of many physical parameters. First of all, the temperatures in the central mass of the reacting material in some of the H-bombs attain values of many kilovolts. This is higher than the presumed values in the center of most stars. We have, for an extremely short time only, of course, situations which are probably hardly duplicated anywhere in the present condition of the universe. The duration of the process is very short — a fraction of a microsecond. Since the disassembly of the crucial part of the material is effected in such times, the material velocities are of the order of 10^8 cm/sec. Both in the fission bombs and the thermonuclear explosions we have, therefore, velocities of material in *bulk* far higher than any otherwise produced under terrestrial conditions. In both the fission bomb explosions and in the thermonuclear ones, an enormous flux of neutrons is present; and, for a short time, one has a condition of a neutron "gas" being present together with the other nuclei at densities comparable to those of ordinary solids. It is possible to conceive an enhancement of these extreme values of physical parameters by special arrangements. It is well known that, by suitable geometrical arrangement of ordinary chemical explosives, one can obtain higher velocities in materials by utilizing the Monroe effect. As is well known, material velocities of the order of tens of kilometers per second were obtained by forming jets initiated by converging explosive shocks. Similar possibilities, no doubt, exist with nuclear explosives.

A great deal of thought and work has gone on in the Los Alamos Laboratory and the Livermore Laboratory on the technological and scientific uses of such concentrated energy sources. It is quite clear, of course, that the availability of such compact and relatively inexpensive ways to produce amounts of energy equal, if necessary, to those obtained by burning millions of tons of coal may find, in the foreseeable future, practical uses, e.g., to mention obvious ones, melting of large amounts of ice and supplying energy in places where transporting ordinary fuels is impracticable or prohibitively costly. In the more remote future, one can imagine the transport and the use of such nuclear explosives to extraterrestrial sites, be it for providing energy sources for maintenance of living conditions or for construction purposes, or fabrication of utilizable materials.

As to more purely scientific applications, several conferences have taken place, at Livermore and at Los Alamos, on such scientific uses. These are suggested by the availability of very high neutrons fluxes and by the possibility of constructing, in a manner accessible to direct observations, conditions which one otherwise finds only near the surface or in the interior of stars.

As mentioned above, for a short time, neutrons are present inside the thermonuclear device in densities such that neutron collisions become rather frequent. For example, collisions between two 14.2 MeV neutrons may lead to the production of neutrons of higher energy, say, above 25 MeV. An indirect measurement of the neutron-neutron collision cross section is therefore possible. Some properties and problems concerning the explosion of stars, e.g., the hydrodynamical questions connected with the behavior of certain novae and supernovae, can be examined during the early stages of the disassembly of the thermonuclear devices. In the latter stages of such explosions, in the behavior of the surrounding material and of the air, subject to high temperatures and pressures, there is still much to be observed which can teach us about conditions like those in the outer regions or on the surface of the sun and other stars. A great number of such observations were proposed and discussed; some of these can be found in the documents summarizing parts of the conferences mentioned above.

Various projects have proposed the use of nuclear sources for the propulsion of space vehicles. One such scheme, described in the next chapter, called Project Orion, originated in Los Alamos and involved the use of nuclear explosions on the outside of the vehicle as a means of propulsion. It would be very desirable to use thermonuclear devices for this purpose both for reasons of economy and the desirability to diminish or possibly eliminate the radioactive contamination attending the fission process and its products.

The technology of the thermonuclear devices does not constitute a finished or complete subject. Improvements are certainly possible in the designs and in arrangements for special purposes. Certainly, the "records" obtained of temperatures, of densities, of velocities of issuing materials, etc., can and will be surpassed in the future. In the numerous possible, peaceful technological and scientific uses, e.g., for removing masses of material (excavations, etc.), for producing reservoirs of thermal energy underground, etc., etc., new designs permitting the release of energy in specific forms will be made. For the problems of rocket propulsion, for example, it would be worthwhile to construct nuclear explosives in such a way that much of the momentum, if not the energy obtained from them, could be released in and confined to given solid angles. For certain other applications in the future, like those of a possible influencing of the flow of air masses, i.e., in the study of methods to be used in "weather control," one will have to consider among other problems the interaction between several, perhaps nearly simultaneous, explosions.

The electronic computing machines have provided for some time now, and will provide even more extensively in the future, the tools for calculating with increasing accuracy both the details of the process of the release of energy and the enormously complicated interactions between the exploding material and its surroundings. Calculations undertaken in order to understand quantitatively the hydrodynamical and radiative processes which are involved have progressed greatly during the years following the first construction of thermonuclear devices. In the near future calculations on computers will help to solve many of the problems concerning three-dimensional hydrodynamical phenomena.

The general theoretical and calculational work directed by Bethe during the early years of Los Alamos will, by following the example he gave of perspicacity and thoroughness, encompass an ever-increasing scope of theoretical understanding and of applied use.

CHAPTER 15

THE ORION PROJECT

FOR A LONG TIME NOW, work has been proceeding on obtaining higher specific impulses with better ratios of payload to the total weight of rockets. One such scheme was known under the name of Project Orion or the Nuclear Pulse Propulsion method.

The first ideas of this sort originated in Los Alamos shortly after the war and the first report on such a scheme, containing preliminary calculations on the technical problems and the possible performance, was written by C.J. Everett and the author in 1955. This is still a classified document, but several descriptions of the Project have appeared in open literature. Shortly after the first Russian Sputnik, T. B. Taylor (formerly of Los Alamos, where he became interested in our proposals) at the General Atomic Laboratory in La Jolla took up this scheme, elaborated it and added new ideas to it. With a group of collaborators, among whom one should mention F. Dyson from the Princeton Institute for Advanced Study, Burton Freeman and Marshall Rosenbluth, a brilliant nucleus of a working group was started. This group grew soon to more than 40 people, which, in addition to physicists, included engineers specializing in structural problems, in electronics, in control systems, etc. A rather impressive amount of work was accomplished in a few years.

The main idea was rather simple. It was to store in the vehicle a large number of nuclear bomblets which would be surrounded by a suitable material. This material, after the explosion of the bomblet, would expand to strike and propel the vehicle *from the outside*. These nuclear charges, together with their surrounding "propellant" mass would be ejected periodically from the vehicle to a given distance under the vehicle. A large fraction of the mass of the propellant would hit a bottom plate on which the vehicle rested and give it successive impulses until the desired velocity of the vehicle was obtained. The original paper dealt with unmanned vehicles. In the planning by the GA group, the bottom plate was connected to the rest of the vehicle by a system of shock absorbers. This would reduce each impulse in producing accelerations to the rest of the vehicles to only several g's. This way the impulses would be entirely tolerable to occupants of the space craft.

Immediately there arose, of course many questions. Would the successive impulses damage the bottom "pusher" plate and render it inoperative by repeated ablations (i.e., wearing away)? Could one provide enough control in the positioning and orientation of the nuclear charges? Would there be an intolerable amount of radiation hitting the inside of the vehicle, etc.? These problems were considered in the original report, which concluded that propulsion of this sort appeared entirely feasible. The immense amount of extremely detailed calculational work was done at the GA Laboratory, and it also lead to the same conclusion. Much of the work had been theoretical but a number of experiments were performed concerning the problem of ablation, the problem of controls, the construction of shock absorbers, structural stability, etc. It appeared that the amount of radiation could be reduced to very tolerable limits.

It should be stressed here that one envisaged vehicles with payloads that are enormous compared to those obtainable by present chemical rockets. One of the smallest designed concerned a vehicle of about one hundred tons gross weight, 10 meters in diameter. This would have a payload comparable to the total gross weight. The specific impulse would be perhaps greater than 2000 seconds. This could be lofted from the ground by a 2-stage Saturn V vehicle. The GA group made detailed designs of vehicles with a payload of 500 tons, with a total weight of perhaps 2500 tons, which could take off from the ground and have a specific impulse of the order of 4000 seconds. In a vehicle of this sort one could think of expeditions to Mars, and perhaps other planets, with a whole group of persons taking part in it.

There is one paramount problem: How can one test and then actually prepare such vehicles when nuclear explosions in the atmosphere and space are banned by the Test Ban Treaty? One answer is of course that the testing of properties of explosions of this type could be done underground. Such work could confirm the non-nuclear experiments showing how much ablation in the pusher plate occurs after it is hit by a gas of desired temperature, density and velocity. When it comes to actual construction, to tests, and to voyages for such vehicles, an international agreement would have to be made, allowing specifically this special type of nuclear explosion for this purpose only. The GA Project had the support of government agencies, and was carried forward under the sponsorship of the United States Air Force, the Advanced Research Projects Agency, and to a modest extent, under NASA.

It is clear that for any really comprehensive and "ambitious" space exploration voyages, very large payloads will be necessary, including perhaps a dozen or more participants, a great number of instruments, tools, provisions, etc. The vehicle could carry its own "taxis" — i.e., small rockets propelled chemically to effect landings on the moon and on the planets, which could then rejoin the parent Orion vehicle.

A number of reports have been written by F. Dyson and others with GA on possible astronomical exploration voyages. (They included visits to some satellites of Jupiter, some of which are particularly suitable for landings made possible by such vehicles!) By now there exists a whole literature, unfortunately still mostly classified, which deals with both the technical problems of construction — e.g., the requirements on the strength, mechanical problems (e.g., vibrations of the pusher plate), the arrangements of instruments, the shielding problems, etc., etc., but also including papers on the variety of missions, discussing both the problem of suitable orbits and the description of the scientific aims of such trips.

Obviously the Project was one of very great magnitude. The cost of constructing the first operational vehicle would be very large. The GA people estimated, perhaps optimistically, that this could be achieved for under one billion dollars. Once one had successful prototypes, the subsequent vehicles would not be enormously expensive. The main cost would be in the nuclear charges, and this would not be prohibitively large.

As mentioned above, the decision whether or not to proceed with the study of these possibilities involves, *inter alia*, the possibility of obtaining some time in the future an international cooperation or at least an agreement for allowing such specified use of nuclear explosives for this exploratory and scientific purpose.

According to some experts (other than just those involved in the conception or work on Orion) the schemes outlined above are among the best so far advanced for construction of really large, maneuverable and powerful space vehicles. No other method known at present gives both the very high thrust and the very high specific impulses which seem obtainable by the nuclear-pulse scheme. It is to be hoped that further theoretical studies and some experimental work will be allowed to proceed in this direction.

Chapter 16

JOHN VON NEUMANN
1903-1957

In John von Neumann's death on February 8, 1957, the world of mathematics lost a most original, penetrating, and versatile mind. Science suffered the loss of a universal intellect and a unique interpreter of mathematics, who could bring the latest (and develop latent) applications of its methods to bear on problems of physics, astronomy, biology, and the new technology. Many eminent voices have described and praised his contributions. It is my aim to add here a brief account of his life and of his work from a background of personal acquaintance and friendship extending over a period of 25 years.

John von Neumann (Johnny, as he was universally known in this country), the eldest of three boys, was born on December 28, 1903, in Budapest, Hungary, at that time part of the Austro-Hungarian empire. His family was well-to-do; his father, Max von Neumann, was a banker. As a small child, he was educated privately. In 1914, at the outbreak of the First World War, he was ten years old and entered the gymnasium.

Budapest, in the period of the two decades around the First World War, proved to be an exceptionally fertile breeding ground for scientific talent. It will be left to historians of science to discover and explain the conditions which catalyzed the emergence of so many brilliant individuals (—their names abound in the annals of mathematics and physics of the present time). Johnny was probably the most brilliant star in this constellation of scientists. When asked about his own opinion on what contributed to this statistically unlikely phenomenon, he would say that it was a coincidence of some cultural factors which he could not make precise: an external pressure on the whole society of this part of Central Europe, a subconscious feeling of extreme insecurity in individuals, and the necessity of producing the unusual or facing extinction. The First World War had shattered the existing economic and social patterns. Budapest, formerly the second capital of the Austro-Hungarian empire, was now the principal town of a small country. It became obvious to many scientists that they would have to emigrate and find a living elsewhere in less restricted and provincial surroundings.

According to Fellner, who was a classmate of his, Johnny's unusual abilities came to the attention of an early teacher (Laslo Ratz). He expressed to Johnny's father the opinion that it would be nonsensical to teach Johnny school mathematics in the conventional way, and they agreed that he should be privately coached in mathematics. Thus, under the guidance of Professor Kürschak and the tutoring of Fekete, then an assistant at the University of Budapest, he learned about the problems of mathematics. When he passed his "matura" in 1921, he was already recognized a professional mathematician. His first paper, a note with Fekete, was composed while he was not yet 18. During the next four years, Johnny was registered at the University of Budapest as a student of mathematics, but he spent most of his time in Zürich at the Eidgenössische Technische Hochschule, where he obtained an undergraduate degree of "Diplomingenieur in Chemie," and in Berlin. He would appear at the end of each semester at the University of Budapest to pass his course examinations (without having attended the courses, which was somewhat irregular). He received his doctorate in mathematics in Budapest at about the same time as his chemistry degree in Zürich. While in Zürich, he spent much of his spare time working on mathematical problems, writing for publication, and corresponding with mathematicians. He had contacts with Weyl and Polya, both of whom were in Zürich. At one time, Weyl left Zürich for a short period, and Johnny took over his course for that period.

It should be noted that, on the whole, precocity in original mathematical work was not uncommon in Europe. Compared to the United States, there seems to be a difference of at least two or three years in specialized education, due perhaps to a more intensive schooling system during the gymnasium and college years. However, Johnny was exceptional even among the youthful prodigies. His original work began even in his student days, and in 1927, he became a Privat-Dozent at the University of Berlin. He held this position for three years until 1929, and during that time, became well-known to the mathematicians of the world through his publications in set theory, algebra, and quantum theory. I remember that in 1927, when he came to Lwów (in Poland) to attend a congress of mathematicians, his work in foundations of mathematics and set theory was already famous. This was already mentioned to us, a group of students, as an example of the work of a youthful genius.

In 1929, he transferred to the University of Hamburg, also as a Privat-Dozent, and in 1930, he came to this country for the first time as a visiting lecturer at Princeton University. I remember Johnny telling me that even though the number of existing and prospective vacancies in German universities was extremely small, most of the two or three score Dozents counted on a professorship in the near future. With his typically rational approach, Johnny computed that the expected number of professorial appointments within three years was three, the number of Dozents was 40! He also felt that the coming political events would make intellectual work very difficult.

He accepted a visiting professorship at Princeton in 1930, lecturing for part of the academic year and returning to Europe in the summers. He became a permanent professor at the University in 1931 and held this position until 1933 when was invited to join the Institute for Advanced Study as a professor, the youngest member of its permanent faculty.

Johnny married Marietta Kovesi in 1930. His daughter, Marina, was born in Princeton in 1935. In the early years of the Institute, a visitor from Europe found a wonderfully informal and yet intense scientific atmosphere. The Institute professors had their offices at Fine Hall (part of Princeton University), and in the Institute and the University departments a galaxy of celebrities was included in what quite possibly constituted one of the greatest concentrations of brains in mathematics and physics at any time and place.

It was upon Johnny's invitation that I visited this country for the first time at the end of 1935. Professor Veblen and his wife were responsible for the pleasant social atmosphere, and I found that the von Neumann's (and Alexander's) houses were the scenes of almost constant gatherings. These were the years of the depression, but the Institute managed to give to a considerable number of both native and visiting mathematicians a relatively carefree existence.

Johnny's first marriage terminated in divorce. In 1938, he remarried during a summer visit to Budapest and brought back to Princeton his second wife, Klara Dan. His home continued to be a gathering place for scientists. His friends will remember the inexhaustible hospitality and the atmosphere of intelligence and wit one found there. Klari von Neumann later became one of the first coders of mathematical problems for electronic computing machines, an art to which she brought some of its early skills.

With the beginning of the war in Europe, Johnny's activities outside the Institute started to multiply. He was performing an enormous amount of work for various scientific projects in and out of the government.

In October, 1954, he was named by presidential appointment a member of the United States Atomic Energy Commission. He left Princeton on a leave of absence and discontinued all commitments with the exception of the chairmanship of the ICBM Committee. Admiral Strauss, chairman of the Commission and a friend of Johnny's for many years, suggested this nomination as soon as a vacancy occurred. Of Johnny's brief period of active service on the Commission, he wrote:

"During the period between the date of his confirmation and the late autumn, 1955, Johnny functioned magnificently. He had the invaluable faculty of being able to take the most difficult problem, separate it into its components, whereupon everything looked brilliantly simple, and all of us wondered why we had not been able to see through to the answer as clearly as it was possible for him to do. In this way, he enormously facilitated the work of the Atomic Energy Commission."

Johnny, whose health had always been excellent, began to look very fatigued in 1954. In the summer of 1955, the first symptoms of a fatal disease were discovered by x-ray examination. A prolonged and cruel illness gradually put an end to all his activities. He died at Walter Reed Hospital in Washington at the age of 53.

Johnny's friends remember him in his characteristic poses: standing before a blackboard or discussing problems at home. Somehow, his gestures, smile, and the expression of the eyes always reflected the kind of thought or the nature of the problem under discussion. He was of middle size, quite slim as a very young man, then increasingly corpulent; moving about in small steps with considerable random acceleration, but never with great speed. A smile flashed on his face whenever a problem exhibited features of a logical or mathematical paradox. Quite independently of his liking for abstract wit, he had a strong appreciation (one might say almost a hunger) for the more earthy type of comedy and humor.

He seemed to combine in his mind several abilities which, if not contradictory, at least seem separately to require such powers of concentration and memory that one very rarely finds them together in one intellect. These are: a feeling for the set-theoretical, formally algebraic basis of mathematical thought, the knowledge and understanding of the substance of classical mathematics in analysis and geometry, and a very acute perception of the potentialities of modern mathematical methods for the formulation of existing and new problems of theoretical physics. All this is specifically demonstrated by his brilliant and original work which covers a very wide spectrum of contemporary scientific thought.

His conversations with friends on scientific subjects could last for hours. There never was a lack of subjects, even when one departed from mathematical topics.

Johnny had a vivid interest in people and delighted in gossip. One often had the feeling that in his memory he was making a collection of human peculiarities as if preparing a statistical study. He followed also the changes brought by the passage of time. When a young man, he mentioned to me several times his belief that the primary mathematical powers decline after the age of about 26, but that a certain more prosaic shrewdness developed by experience manages to compensate for this gradual loss, at least for a time. Later, this limiting age was slowly raised.

He engaged occasionally in conversational evaluations of other scientists; he was, on the whole, quite generous in his opinions, but often able to damn by faint praise. The expressed judgment was, in general, very cautious, and he was certainly unwilling to state any final opinions about others: "Let Rhadamantys and Minos ... judge ...." Once when asked, he said that he would consider Erhard Schmidt and Hermann Weyl among the mathematicians who especially influenced him technically in his early life.

Johnny was regarded by many as an excellent chairman of committees (this peculiar contemporary activity). He would press strongly his technical views, but defer rather easily on personal or organizational matters.

In spite of his great powers and his full consciousness of them, he lacked a certain self-confidence, admiring greatly a few mathematicians and physicists who possessed qualities which he did not believe he himself had in the highest possible degree. The qualities which evoked this feeling on his part were, I felt, relatively simple-minded powers of intuition of new truths, or the gift for a seemingly irrational perception of proofs or formulation of new theorems.

Quite aware that the criteria of value in mathematical work are, to some extent, purely aesthetic, he once expressed an apprehension that the values put on abstract scientific achievement in our present civilization might diminish: "The interests of humanity may change, the present curiosities in science may cease, and entirely different things may occupy the human mind in the future." One conversation centered on the ever accelerating progress of technology and changes in the mode of human life, which gives the appearance of approaching some essential singularity in the history of the race beyond which human affairs, as we know them, could not continue.

His friends enjoyed his great sense of humor. Among fellow scientists, he could make illuminating, often ironical, comments on historical or social phenomena with a mathematician's formulation, exhibiting the humor inherent in some statement true only in the vacuous set. These often could be appreciated only by mathematicians. He certainly did not consider mathematics sacrosanct. I remember a discussion in Los Alamos, in connection with some physical problems where a mathematical argument used the existence of ergodic transformations and fixed points. He remarked with a sudden smile, "Modern mathematics can be applied after all! It isn't clear, *a priori*, is it, that it could be so . . . ."

I would say that his main interest after science was in the study of history. His knowledge of ancient history was unbelievably detailed. He remembered, for instance, all the anecdotical material in Gibbon's Decline and Fall and liked to engage after dinner in historical discussions. On a trip south, to a meeting of the American Mathematical Society at Duke University, passing near the battlefields of the Civil War he amazed us by his familiarity with the minutest features of the battles. This encyclopedic knowledge molded his views on the course of future events by inducing a sort of analytic continuation. I can testify that in his forecasts of political events leading to the Second World War and of military events during the war, most of his guesses were amazingly correct. After the end of the Second World War, however, his apprehensions of an almost immediate subsequent calamity, which he considered as extremely likely, proved fortunately wrong. There was perhaps an inclination to take a too exclusively rational point of view about the cases of historical events. This tendency was possibly due to an over-formalized game theory approach.

Among other accomplishments, Johnny was an excellent linguist. He remembered his school Latin and Greek remarkably well. In addition to English, he spoke German and French fluently. His lectures in this country were well known for their literary quality (with very few characteristic mispronunciations which his friends anticipated joyfully, e.g., "integhers"). During his frequent visits to Los Alamos and Santa Fe (New Mexico), he displayed a less perfect knowledge of Spanish, and on a trip to Mexico, he tried to make himself understood by using "neo-Castilian," a creation of his own — English words with an "el" prefix and appropriate Spanish endings.

Before the war, Johnny spent the summers in Europe on vacations and lecturing (in 1935 at Cambridge University, in 1936 at the Institut Henri Poincaré in Paris). Often he mentioned that personally he found doing scientific work there almost impossible because of the atmosphere of political tension. After the war he undertook trips abroad only unwillingly.

Ever since he came to the United States, he expressed his appreciation of the opportunities here and very high hopes for the future of scientific work in this country.

To follow chronologically von Neumann's interests and accomplishments is to review a large part of the whole scientific development of the last three decades. In his youthful work, he was concerned not only with mathematical logic and the axiomatics of set theory, but, simultaneously, with the substance of set theory itself, obtaining interesting results in measure theory and the theory of real variables. It was in this period also that he began his classical work on quantum theory, the mathematical foundation of the theory of measurement in quantum theory and the new statistical mechanics. His profound studies of operators in Hilbert spaces also date from this period. He pushed far beyond the immediate needs of physical theories, and initiated a detailed study of rings of operators, which has independent mathematical interest. The beginning of the work on continuous geometries belongs to this period as well.

Von Neumann's awareness of results obtained by other mathematicians and the inherent possibilities which they offer is astonishing. Early in his work, a paper by Borel on the minimax property led him to develop in the paper, *Zur Theorie der Gesellschaft-Spiele,* ideas which culminated later in one of his most original creations, the theory of games. An idea of Koopman on the possibilities of treating problems of classical mechanics by means of operators on a function space stimulated him to give the first mathematically rigorous proof of an ergodic theorem. Haar's construction of measure in groups provided the inspiration for his wonderful partial solution of Hilbert's fifth problem, in which he proved the possibility of introducing analytical parameters in compact groups.

In the middle 30's, Johnny was fascinated by the problem of hydrodynamical turbulence. It was then that he became aware of the mysteries underlying the subject of non-linear partial differential equations. His work, from the beginning of the Second World War, concerns a study of the equations of hydrodynamics and the theory of shocks. The phenomena described by these non-linear equations are baffling analytically and defy even qualitative insight by present methods. Numerical work seemed to him the most promising way to obtain a feeling for the behavior of such systems. This impelled him to study new possibilities of computation on electronic machines, *ab initio.* He began to work on the theory of computing and planned the work, to remain unfinished, on the theory of automata. It was at the outset of such studies that his interest in the working of the nervous system and the schematized properties of organisms claimed so much of his attention.

This journey through many fields of mathematical sciences was not a result of restlessness. Neither was it a search for novelty, nor a desire for applying a small set of general methods to many diverse special cases. Mathematics, in contrast to theoretical physics, is not confined to a few central problems. The search for unity, if pursued on a purely formal basis, von Neumann considered doomed to failure. This wide range of curiosity had its basis in some metamathematical motivations and was influenced strongly by the world of physical phenomena — these will probably defy formalization for a long time to come.

Mathematicians, at the outset of their creative work, are often confronted by two conflicting motivations: the first is to contribute to the edifice of existing work — it is there that one can be sure of gaining recognition quickly by solving outstanding problems — the second is the desire to blaze new trails and to create new syntheses. This latter course is a more risky undertaking, the final judgment of value or success appearing only in the future. In his early work, Johnny chose the first of these alternatives. It was toward the end of his life that he felt sure enough of himself to engage freely and yet painstakingly in the creation of a possible new mathematical discipline. This was to be a combinatorial theory of automata and organisms. His illness and premature death permitted him to make only a beginning.

In his constant search for applicability and in his general mathematical instinct for all exact sciences, he brought to mind Euler, Poincaré, or in more recent times, perhaps Hermann Weyl. One should remember that the diversity and complexity of contemporary problems surpass enormously the situation confronting the first two named. In one of his last articles, Johnny deplored the fact that it does not seem possible nowadays for any one brain to have more than a passing knowledge of more than one-third of the field of pure mathematics.

Early work, set theory, algebra

The first paper, a joint work with Fekete, deals with zeros of certain minimal polynomials. It concerns a generalization of Fejér's theorem on location of the roots of Tchebyscheff polynomials. Its date is 1922. Von Neumann was not quite eighteen when the article appeared.

Another youthful work is contained in a paper (in Hungarian with a German summary) on uniformly dense sequences of numbers. It contains a theorem on the possibility of re-ordering dense sequences of points so they will become uniformly dense; the work does not yet indicate the future depth of formulations nor is it technically difficult, but the choice of subject and the conciseness of technique in proofs begins to indicate the combination of set-theoretical intuition and the algebraic technique of his future investigations.

The set-theoretical orientation in the thinking of a great number of young mathematicians is quite characteristic of this era. The great ideas of George Cantor, which found their fruition finally in the theory of real variables, topology and later in analysis, through the work of the great Frenchmen, Baire, Borel, Lebesgue, and others, were not yet commonly part of the fundamental intuitions of young mathematicians at the turn of the century. After the end of the First World War, however, one notices that these ideas became more, as it were, naturally instinctive for the new generation.

His second paper in the Acta Szeged on transfinite ordinals already shows von Neumann in his characteristic form and style in dealing with the algebraic treatment of set theory. The first sentence states frankly: "The aim of this work is to formulate concretely and precisely the idea of Cantor's ordinal numbers." As the preface states, the heretofore somewhat vague formulation of Cantor himself is replaced by definitions which can be given in the system of axioms of Zermelo. Moreover, a rigorous foundation for the definition by transfinite induction is outlined. The introduction stresses the strictly formalistic approach, and von Neumann states somewhat proudly that the symbols ". . ." (for "et cetera") and similar expressions are never employed. This treatment of ordinal numbers, later also considered by Kuratowski, is to this day the best introduction of this idea, so important for "constructions" in abstract set theory. Each ordinal number by von Neumann's definition is the set of all smaller ordinal numbers. This leads to a most elegant theory and moreover allows one to avoid the concept of ordertype, which is vague insofar as the set of all ordered sets similar to a given one does not exist in axiomatic set theory.

His paper on Prüfer's theory of ideal algebraic numbers begins to indicate his future breadth of interests. The paper deals with set theoretical questions and enumeration problems about relatively prime ideal components. Prüfer had introduced ideal numbers as "ideal solutions of infinite systems of congruences." Von Neumann starts with methods analogous to Kürschak and Bauer's work on Hensel's *p*-adic numbers. Here again, von Neumann exhibits the techniques which were to become so prevalent in the following decades in mathematical research — of continuing algebraical constructions, originally considered on finite sets, to the domain of the infinitely enumerable and the continuum. Another indication of his algebraic interests is a short note on Minkowski's theory of linear forms.

A desire to axiomatize — and this in a sense more formal and precise than that originally considered by logicians at the beginning of the 20th century — shows through much of the early work. From around 1925 to 1929, most of von Neumann's papers deal with attempts to spread the spirit of axiomatization even through physical theory. Not satisfied with the existing formulations, even in set theory itself, he states again quite frankly in the first sentence of his paper on the axiomatization of set theory: "The aim of the present work is to give a logically unobjectionable axiomatic treatment of set theory"; the next sentence reads, "I would like to say something at first about difficulties which make such a construction of set theory desirable."

The last sentence of this 1925 paper is most interesting. Von Neumann points out the limits of any axiomatic formulation. There is here perhaps a vague forecast of Gödel's results on the existence of undecidable propositions in any formal system. The concluding sentence is: "We cannot, for the present, do more than to state that there are here objections against set theory itself, and there is no way known at present to avoid these difficulties." (One is reminded here, perhaps, of an analogous statement in an entirely different domain of science: Pauli's evaluation of the state of relativistic quantum theory written in his Handbuch der Physik article and the still mysterious role of infinities and divergences in field theory.)

His second paper on this subject has the title, *The axiomatization of set theory* (*An axiomatization of set theory* was the 1925 title).

The conciseness of the system of axioms is surprising, the introduction of objects of the first and second type corresponding, respectively, to sets and properties of sets in the naïve set theory; the axioms take only a little more than one page of print. This is sufficient to build up practically all of the naïve set theory and therewith all of modern mathematics and constitutes, to this day, one of the best foundations for set-theoretical mathematics. Gödel, in his great work on the independence of the axiom of choice, and on the continuum hypothesis, uses a system inspired by this treatment. It is noteworthy that in his first paper on the axiomatization of set theory, von Neumann recognizes explicitly the two fundamentally different directions taken by mathematicians in order to avoid the antinomies of Burali-Forti, Richard and Russell. One group, containing Russell, J. König, Brouwer, and Weyl, takes the more radical point of view that the entire logical foundations of exact sciences have to be restricted in order to prevent the appearance of paradoxes of the above type. Von Neumann says, "the general impression of their activity is almost crushing." He objects to Russell's building the system of mathematics on the highly problematic axiom of reducibility, and objects to Weyl's and Brouwer's rejection of what he considers as the greater part of mathematics and set theory.

He has more sympathy with the second less radical group, naming in it Zermelo, Fraenkel, and Schoenflies. He considers their work, *including his own*, as far from complete, stating explicitly that the axioms appear somewhat arbitrary. He states that one cannot show in this fashion that antinomies are really excluded but while naïve set theory cannot be considered too seriously, at least much of what it contains can be rehabilitated as a formal system, and the sense of "formalistic" can be defined in a clear fashion.

Von Neumann's system gives the first foundation of set theory on the basis of a finite number of axioms of the same simple logical structure as have, e.g., the axioms of elementary geometry. The conciseness of the system of axioms and the formal character of the reasoning employed seem to realize Hilbert's goal of treating mathematics as a finite game. Here one can divine the germ of von Neumann's future interest in computing machines and the "mechanization" of proofs.

Starting with the axioms, the efficiency of the algebraic manipulation in the derivation of most of the important notions of set theory is astounding; the economy of the treatment seems to indicate a more fundamental interest in brevity than in virtuosity for its own sake. It thereby helped prepare the grounds for an investigation of the limits of finite formalism by means of the concept of "machine" or "automaton."

It seems curious to me that in the many mathematical conversations on topics belonging to set theory and allied fields, von Neumann even seemed to think formally. Most mathematicians, when discussing problems in these fields, seemingly have an intuitive framework based on geometrical or almost tactile pictures of abstract sets, transformations, etc. Von Neumann gave the impression of operating sequentially by purely formal deductions. What I mean to say is that the basis of his intuition, which could produce new theorems and proofs just as well as the "naïve" intuition, seemed to be of a type that is much rarer. If one has to divide mathematicians, as Poincaré proposed, into two types — those with visual and those with auditory intuition — Johnny perhaps belonged to the latter. In him, the "auditory sense," however, probably was very abstract. It involved, rather, a complementarity between the formal appearance of a collection of symbols and the game played with them on the one hand, and an interpretation of their meanings on the other. The foregoing distinction is somewhat like that between a mental picture of a physical chess board and a mental picture of a sequence of moves on it, written down in algebraic notation.

In conversations, on the present status of foundations of mathematics, von Neumann seemed to imply that in his view, the story is far from having been told. Gödel's discovery should lead to a new approach to the understanding of the role of formalism in mathematics, rather than be considered as closing the subject.

A later paper translates into strictly axiomatic treatment what was done informally in his second paper. The first part of the paper deals with the introduction of the fundamental operations in set theory, the foundation of the theories of equivalence, similarity, well-ordering, and finally, a proof of the possibility of definition by finite or transfinite induction, including a treatment of ordinal numbers. Von Neumann rightly insists at the end of his introduction to the paper that transfinite induction was not rigorously introduced before in any axiomatic or non-axiomatic system of set theory.

Perhaps the most interesting of von Neumann's papers on axiomatics of set theory is his 1929 paper. It has to do with a certain necessary and sufficient condition which a property of sets must satisfy in order to define a set of sets. The condition is that there must not exist a one-to-one correspondence between all sets and the sets which have the property in question. This existential principle for sets had been assumed as an axiom by von Neumann and some of the axioms assumed in other systems, in particular the axiom of choice, had been derived from it. Now it is shown that, vice versa, these other axioms imply von Neumann's axiom, which thereby is proved consistent, provided the usual axioms are.

His great paper in the Mathematische Zeitschrift, *Zur Hilbert-schen Beweistheorie,* is devoted to the problem of the freedom from contradiction of mathematics. This classical study contains an exposition of the primitive ideas underlying mathematical formalisms in general. It is stressed that the whole complex of problems, originated and developed by Hilbert and also treated by Bernays and Ackermann, have not been satisfactorily solved. In particular, it is pointed out that Ackermann's proof of freedom from contradiction cannot be carried through for classical analysis. It is replaced by a rigorous finitary proof for a certain subsystem. In fact von Neumann's proof shows (although this is not stated explicitly) that finitely iterated application of quantifiers and propositional connectives to any finitary (i.e., decidable) relations is consistent. This is not far from the limit of what can be obtained on the basis of Hilbert's original program, i.e., with strictly finitary methods. But von Neumann at that time conjectured that all of analysis can be proved consistent with the same method. At the present time, one cannot escape the impression that the ideas initiated by the work of Hilbert and his school, developed with such precision, and then revolutionized by Gödel, are not yet exhausted. It might be that we are in the midst of another great evolution: the "naïve" treatment of set theory and the formal metamathematical attempts to contain the set of our intuitions about infinity are, I think, turning toward a future "super set theory." Several times in the history of mathematics, the intuitions or, one might better say, common vague feelings of leading mathematicians about problems of existing science, later became incorporated in a formal "super system" dealing with the essence of problems in the original system.

Von Neumann pursued his interest in problems of foundations of mathematics until the end of his life. A quarter of a century after the appearance of the above series of papers, one can see the imprint of this work in his discoveries in the plans for the logic of computing machines.

Parallel to the work on foundations of mathematics, there come specific results in set theory itself and set-theoretically motivated theorems in real variables and in algebra. For example, von Neumann shows the existence of a set M of real numbers, of the power of the continuum, such that any finite number of the elements of M are algebraically independent. The proof is given effectively without the axiom of choice. In a paper in Fundamenta Mathematicae, the same year, a decomposition of the interval is given into countably many disjoint and congruent subsets. This solved a problem of Steinhaus — a special ingenuity is required to have such a decomposition on an interval — the corresponding construction of Hausdorff for the circumference of a circle is much easier. (This is due to the fact that the circumference of a circle may be regarded as a group manifold.)

In his paper on the general measure theory, in Fundamenta (1928), the problem of a finitely additive measure is treated for subsets of groups. The paradoxical decompositions of the sphere by Hausdorff and the wonderfully strange decompositions of Banach and Tarski are generalized from the Euclidean space to general non-Abelian groups. The affirmative results of Banach on the possibility of a measure for *all* subsets of the plane are generalized to the case of subsets of a commutative group. The final conclusion is that all solvable groups are "measurable" (i.e. such measure can be introduced in them).

The problems and methods of this article form one of the first instances of a trend which developed strongly since that time, that of generalizing the set-theoretical results from Euclidean space to more general topological and algebraic structures. The "congruence" of two sets is understood to mean equivalence under a transformation belonging to a given group of transformations. The measure is a general additive set function. Again, the formulation of the problem presages the work of Haar and the study of Hausdorff-Banach-Tarski paradoxical decompositions.

In the same *"annus mirabilis,"* 1928, there appears the article on the theory of games. This is his first work on what was to become later an important combinatorial theory with so many applications and developments vigorously continuing at the present time. It is hard to believe that beginning with 1927, simultaneously with the work discussed above, he could have published numerous papers on the mathematical foundations of quantum theory, probability in statistical quantum theory, and some important results on representation of continuous groups!

Theory of functions of real variables, measure theory, topology, continuous groups

We shall briefly mention some of his results in this field viewed against the background of his other work.

A 1931 paper solves a problem of Haar. It concerns the

selection of representatives from classes of functions which are equivalent up to a set of measure zero from linear manifolds over products of powers of finite systems. The problem is generalized to measures other than Lebesgue's and an analogous problem is solved affirmatively.

The next year he proved an important fact in measure theory: Any Boolean mapping between two classes of measurable sets (on two measure spaces) which preserves their measures is generated by a point transformation preserving measure. This result is important in showing the equivalence of rather general measure spaces, when they are separable and complete, to Euclidean spaces with Lebesgue measure, and permits one to reduce the study of Boolean algebras of measurable sets to ordinary measures.

In 1934 von Neumann proves the uniqueness of the Haar measure as constructed by A. Haar, if one requires either left or right invariance of the (Lebesgue-type) measure under group multiplication. The theorem on uniqueness is proved for compact groups. A construction different from that of Haar is employed to introduce his measure. This paper precedes the construction of a general theory of almost periodic functions on separable topological groups and allows a theory of their orthogonal representations.

In a joint paper with P. Jordan a solution is given to a question raised by Fréchet of the characterization of generalized Hilbert space among linear metric spaces. The condition which is necessary and sufficient, strengthening a result of Fréchet, is: A linear metric space L is isometric with a Hilbert space if and only if every two-dimensional linear subspace is isometric with a Euclidean space.

In the Russian Sbornik, von Neumann deals again with the problem of the uniqueness of Haar's measure. The previous proof of uniqueness was accomplished through a constructive process different from that of Haar, which contained no arbitrary elements and led automatically to the uniqueness of the measure. In this paper an independent treatment of uniqueness of the left- and right-invariant exterior measure is given for locally compact separable groups. (A different proof was obtained simultaneously by André Weil.)

In a joint paper with Kuratowski, precise and strong results are obtained on the projectivity of certain sets of real numbers defined by transfinite induction. The celebrated set of Lebesgue, shown previously by Kuratowski to be of projective class 3, is shown to be a difference of two analytic sets and therefore of the second projective class. A general theorem is proved on the analytic character of sets (in the sense of Hausdorff) obtained by certain general constructions. This result is likely to play an important role in the still incomplete theory of projective sets.

The Memoire in Compositio Mathematica, on infinite direct products contains an algebraic theory of operators and a measure theory for such systems, so important in modern abstract analysis. It summarizes some of the previous work on the algebra of functional operators and topology of rings of operators, including the non-separable hyper-Hilbert spaces. Methodologically and in the actual constructions, this paper is both a forerunner of and a good introduction to much of the recent work in mathematics dealing, so to say, with the pyramiding of algebraical notions. Starting with a vector space, one deals first with their products, then with linear operators on these structures; and finally with classes of such operators whose algebraical properties are investigated again "on the first level." Von Neumann intended to discuss the analogy of this elaborate system with the theory of hyperquantization in quantum theory, and considered the paper in particular as a mathematical preparation for dealing with non-enumerable products.

The 1929 paper is, I believe, the first one in which a very significant contribution is made to the complex of questions originating in Hilbert's fifth problem: the possibility of a change of parameters in a continuous group so that the group operation will become analytic. The work deals with subgroups of the group of linear transformations of n-dimensional space and the result is affirmative: Every such continuous group has a normal subgroup, locally representable analytically and in a one-to-one way by a finite number of parameters.

This is the first of the theorems showing that the group property prevents the "pathological" possibilities common in the theory of functions of a real variable. The results of the paper, later generalized and simplified by E. Cartan for subgroups of general Lie groups, give detailed insight into the structure of such groups by the representation of elements as products of exponential operators. They show that every linear manifold which contains with every two matrices U, V also their commutator $UV - VU$, is an infinitesimal group of an entire group G. This paper is historically important, as preceding the work of Cartan, a later paper of Ado, and of course, von Neumann's own paper where Hilbert's fifth problem is solved for compact groups.

This celebrated result is based on and stimulated by a paper of Haar (in the same volume of Annals of Math.) where an invariant measure function is introduced in continuous groups. Von Neumann shows, (using an analogue of the Peter-Weyl integration on groups and employing the theorem on approximability of functions by linear combinations of a finite number of eigenfunctions of an integral operator—the method of E. Schmidt's dissertation—and with an ingenious use of Brouwer's theorem on invariance of region in Euclidean n-dimensional space)—that every compact and n-dimensional topological group is continuously isomorphic to a closed group of unitary matrices of a finite dimensional Euclidean space.

The method of this article allows one to represent more general (not necessarily n-dimensional) groups as subgroups of infinite products of such n-dimensional groups. In the second part of the paper, an example is given of a finite dimensional non-compact group of transformations acting on Euclidean space in such a way that no change of parameters in the space will make the given transformations analytic. It was almost twenty years before the solution of Hilbert's fifth problem was completed, to include the "open" (i.e., non-compact) n-dimensional groups, by the work of Montgomery and Gleason. Von Neumann's achievement required an intimate knowledge of both the set-theoretical, real variable techniques, a feeling for the spirit of Brouwerian topology, and a real understanding of the technique of integral equations and the calculus of matrices.

A combination of virtuosity in the mode of abstract algebraic thinking and the employment of analytical techniques can be seen in the joint paper with Jordan and Wigner on an algebraic generalization of the quantum mechanical formalism. This is conceived as a possible starting point for future generalizations of the quantum mechanical theories and deals with commutative but not associative hypercomplex algebras. The essential result is that all such formally real finite and commutative r-number systems are merely matrix algebras, with one exception. This exception, however, seems too narrow for the generalizations needed in quantum theories.

Hilbert space, operator theory, rings of operators

His first interest in this subject also stemmed from work on rigorous formulations of quantum theory. In 1954, in a questionnaire which von Neumann answered for the National Academy of Sciences, he named this work as one of his three contributions to mathematics that he considered most important. In sheer bulk alone, papers on these subjects comprise roughly one-third of his printed work. These contain a very detailed analysis of properties of linear operators and an algebraical study of classes (rings) of operators in infinite-dimensional spaces. The result fulfills his avowed purpose stated in the book, *Mathematische Grundlagen der Quantenmechanik* of demonstrating that the ideas originally introduced by Hilbert are capable of constituting an adequate basis for the physical considerations of quantum theory, and that no need exists for the introduction of new mathematical schemes for these physical theories. Von Neumann's unbelievably detailed and meticulous work of classification of the properties of linearity for unitary spaces resolves many problems for unbounded operators. It gives a complete theory of hypermaximal transformations and brings Hilbert space almost as completely within the grasp of the mathematician as is the case with the finite dimensional Euclidean space.

His interest in this subject was continuous throughout his scientific life. Even up to the end, in the midst of work on other subjects, he obtained and published results on the properties of operators and spectral theory. No one has done more than von Neumann, at least in the unitary case and for linear transformations, toward the resolution of the mysteries of non-compactness. Future work in this direction will be based on his results for a long time to come. This work is now being vigorously continued by, among others, his collaborators and former students — Murray in particular — and one is entitled to expect from them further valuable insight into the properties of linear operators.

Theory of lattices, continuous geometry

Here again, von Neumann's interest was stimulated by the possibility of applying these new combinatorial and algebraic schemes to quantum theory. Lattice theory, around 1935, was being developed and generalized by Garrett Birkhoff from the original formulations of Dedekind. At about the same time, an algebraic and set-theoretical study of Boolean algebras was systematically undertaken by M. H. Stone. I remember that in the summer of 1935, Birkhoff, Stone, and von Neumann, on their way from a mathematical meeting in Moscow, stopped in Warsaw and presented short talks at a meeting of the Warsaw Mathematical Society on the new developments in these fields with novel formulations of the logic of quantum theory. The ensuing discussions led one to expect far-reaching applications of the general Boolean algebra and lattice theory formulations of the language of quantum theory. Von Neumann returned to these attempts several times later in his work, but most of his thoughts in this direction are in unpublished notes.

His work on continuous geometries and geometries without points was motivated by the belief that the primitive notions of quantum theory deal with such entities; obviously, the "universe of discourse" consists of certain classes of identified points or linear manifolds in Hilbert space. (This is noted explicitly by Dirac in his book.)

Some of this work was considered for presentation in colloquium lectures; an account of it is contained in the Princeton Institute Lectures; some remains in manuscript form. In conversations with him touching upon these problems, my impression was that, beginning about 1938, von Neumann felt that the new facts and problems of nuclear physics gave rise to problems of an entirely different type and made it less important to insist on a mathematically flawless formulation of a quantum theory of atomic phenomena alone. After the end of the war, he would express sentiments, somewhat similar to remarks reportedly made by Einstein, that the bewildering wealth of nuclear and elementary particle physics make premature any attempt to formulate a general quantum theory of fields, at least for the time being.

Theoretical physics

In the questionnaire for the National Academy of Science mentioned earlier, von Neumann selected as his most important scientific contributions work on mathematical foundations of quantum theory and the ergodic theorem (in addition to the theory of operators discussed above). This choice, or rather restriction, might appear curious to most mathematicians, but is psychologically interesting. It seems to indicate that perhaps his main desire and one of his strongest motivations was to help re-establish the role of mathematics on a *conceptual level* in theoretical physics. The drifting apart of abstract mathematical research and of the main stream of ideas in theoretical physics since the end of the First World War is undeniable. Von Neumann often expressed concern that mathematics might not keep abreast of the exponential increase of problems and ideas in physical sciences. I remember a conversation in which I advanced the fear that a sort of Malthusian divergence may take place: the physical sciences and technology increase in a geometrical ratio and mathematics in an arithmetical progression.

He said that this indeed *might* be the case. Later in the discussion, we both managed to cling, however, to the hope that the mathematical method would remain for a long time in conceptual control of the exact sciences!

One of his early papers was published jointly with Hilbert and Nordheim. According to the preface it is based on a lecture given by Hilbert in the winter of 1926 on the new developments in quantum theory, and prepared with the help of Nordheim. According to the introduction, important parts of the mathematical formulation and discussion are due to von Neumann.

The stated aim of the paper is to introduce, instead of strictly functional relationships of classical mechanics, probability relationships. It also formulates the ideas of Jordan and Dirac in a considerably simpler and more comprehensible manner. Even now, 30 years later, it is difficult to overestimate the historical importance and influence of this paper and the subsequent work of von Neumann in this direction. The great program of Hilbert in axiomatization gains here another vital domain of application, an isomorphism between a physical theory and the corresponding mathematical system. An explicit statement in the introduction to the paper is that it is difficult to understand the theory if its formalism and its physical interpretation are not separated concisely and completely. Such separation is the aim of the paper, even though it is admitted that a complete axiomatization was at the time impossible. The paper contains an outline of the calculus of operators which correspond to physical observables, discusses the properties of Hermitean operators, and altogether forms a prelude to the *Mathematische Begründung der Quantenmechanik*.

At least two mathematical contributions are of importance in the history of quantum mechanics: The mathematical treatment by Dirac did not always satisfy the requirements of mathematical rigor. For example, it operated with the assumption that every self-adjoint operator can be brought into diagonal form, which forced one to introduce for those operators where this cannot be done, the famous "improper" functions of Dirac. *A priori* it might seem, as von Neumann states, that just as

Newtonian mechanics required (at that time) the contradictory infinitesimal calculus, so quantum theory seemed to need a new form of analysis of infinitely many variables. Von Neumann's achievement was to show that this was not the case, namely, that the transformation theory could be put on a clear mathematical basis not by making precise the methods of Dirac but by developing Hilbert's spectral theory of operators. In particular, this was accomplished by his study of non-bounded operators going beyond the classical theory of Hilbert, F. Riesz, E. Schmidt, and others.

The second contribution forms the substance of Chapters 5 and 6 of his book. It has to do with the problems of measure and reversibility in quantum theory. Almost from the beginning, when the ideas of Heisenberg, Schrödinger, Dirac, and Born were enjoying their first sensational success, questions were raised on the role of indeterminism in the theory and proposals made to explain it by the assumption of possible "hidden" parameters which, when discovered in the future, would allow a return to a more deterministic description. Von Neumann shows that the statistical character of statements of the theory is not due to the fact that the state of the observer who performs the measurement is unknown. The system comprising both the observed and observer leads to the uncertainty relations even if one admits an exact state of the observer. This is shown to be the consequence of the previous assumptions of quantum theory involving the general properties of association of physical quantities with operators in Hilbert space.

Apart from the great didactic value of this work which presented the ideas of the new quantum theory in a form congenial and technically interesting to mathematicians, it is a contribution of absolutely first importance, considered as an attempt to make a rational presentation of a physical theory which, as originally conceived by the physicists, was based on non-universally communicable intuitions. While it cannot be asserted that it introduced ideas of novel physical import — and the quantum theory as conceived during these years by Schrödinger, Heisenberg, Dirac, and others still forms only an incomplete theoretical skeleton for the more baffling physical phenomena discovered since — von Neumann's treatment allows at least *one* logically and mathematically clear basis for a rigorous treatment.

Analysis, numerical work, work in hydrodynamics

In 1930 Von Neumann proved a fundamental lemma in the calculus of variations due to Radó. He proved it by means of a simple geometrical construction, (the lemma asserts that a function $z = f(x,y)$ satisfies a Lipschitz condition with a constant Δ if no plane whose maximal inclination is greater than Δ meets the boundary of the surface defined by the given function in three or more points). The paper is also interesting in that the method of proof involves direct geometric visualizations somewhat rare in von Neumann's published work.

A later paper contains one of the impressive achievements of mathematical analysis in the last quarter century. It is the first precise mathematical result in a whole field of investigation: a rigorous treatment of the ergodic hypothesis in statistical mechanics. It was stimulated by the discovery by Koopman of the possibility of reducing the study of Hamiltonian dynamical systems to that of operators in Hilbert space. Using Koopman's representation, von Neumann proved what is now known as the weak ergodic theorem, or the convergence in measure of the means of functions of the iterated, measure-preserving transformation on a measure space. It is this theorem, strengthened shortly afterward by G. D. Birkhoff, in the form of convergence almost everywhere, which provided the first rigorous mathematical basis for the foundations of classical statistical mechanics. The subsequent developments in this field and the numerous generalizations of these results are well-known and will not be mentioned here in detail. Again, this success was due to the combination of von Neumann's mastery of the techniques of the set-theoretically inspired methods of analysis and those originating in his own work on operators on Hilbert space. Still another domain of mathematical physics became accessible to precise and general considerations of modern analysis. In this instance again, a great initial advance was scored, but, of course, here the story is really quite unfinished; a mathematical treatment of the foundations of statistical mechanics, in the case of classical dynamics, is far from complete! It is very well to have the ergodic theorems and the knowledge of the existence of metrically transitive transformations; these facts, however, form only a basis of the subject. Von Neumann often expressed in conversations a feeling that future progress will depend on theorems which would allow a mathematically satisfactory treatment of the subsequent parts of the subject. A complete mathematical theory of the Boltzmann equation and precise theorems on the rates at which systems tend toward equilibrium are needed.

Important work is contained in a joint work with S. Bochner. The use of operator-theoretical methods allows a rather profound discussion of the properties of partial differential equations of the type $A\phi = \partial\phi/\partial t$, $\phi = \phi(t; x, y, z)$, with A of the form

$$A = a\left(\frac{\partial^2}{\partial x^2} + \frac{\partial^2}{\partial y^2} + \frac{\partial^2}{\partial z^2}\right)$$

as in problems of heat conduction, or $A = (2\pi i/h)H$, where H is the energy operator in Schrödinger's quantum mechanical equation for non-stationary states.

An example of the combination of analytical and geometrical techniques is the joint work with Schoenberg. If S is a metric space, $d(f, g)$ being the distance between any two elements of it, we call a function, f_t, whose values lie in S and which is continuous, a screw function if $d(f_t, f_s) = F(t - s)$. The fundamental theorem determines the class of all such functions on a Hilbert space and determines their form. (Any such function $F(t)$ is given by

$$F^2(t) = \int_0^\infty \frac{\sin^2 tu}{u^2} \, d\gamma(u)$$

where $\gamma(u)$ is non-decreasing for $u \geq 0$ and such that $\int_1^\infty u^{-2} d\gamma(u)$ exists.)

In applied analysis, the war years brought a need for quick estimates and approximate results in problems which often do not present a very "clean" appearance, that is to say, are mathematically very inhomogeneous, the physical phenomena to be calculated involving, in addition to the main process, a number of external perturbations whose effect cannot be neglected or even separated in additional variables. This situation comes up often in questions of present day technology and forces one, at least initially, to resort to numerical methods, not because one requires the results with high accuracy but simply to achieve qualitative orientation! This fact, perhaps somewhat deplorable for a mathematical purist, was realized by von Neumann whose interest in numerical analysis increased greatly at that time.

A joint work with H. H. Goldstine, presents a study of the problem of the numerical inversion of matrices of high order. Among other things, it attempts to give rigorous error estimates. Interesting results are obtained on the precision achievable in inverting matrices of order ~ 150. Estimates are obtained "in the general case." ("General" means that under plausible assumed statistics, these estimates hold with the exception of a set of low probability.)

In a subsequent paper on this subject, the problem is reconsidered in an effort to obtain *optimum* numerical estimates. Given a matrix $A = (a_{ij})(i,j = 1,2, \cdots n)$ whose elements are independent random variables, each normally distributed, the probability that the upper bound of this matrix exceeds $2.72\sigma n^{1/2}$ where σ is the dispersion of each variable, is less than $.027 \times 2^{-n} n^{-1/2}$.

The development of the fast electronic computing machines was prompted primarily by the need of a quick orientation and answer to problems in mathematical physics and engineering. There is, as a byproduct, an opportunity for some lighter work! Thus, for example, one can now try to satisfy, to a modest extent, some of the curiosity which is felt about certain interesting sequences of integers, e.g., to mention the simplest ones, the frequency of the sequence of digits in the development of

e and π, carried to many thousands of places. One such computation, performed on the machine at the Institute for Advanced Study, gives the first 2,000 partial quotients of the cube root of 2 in its development as a continued fraction. Johnny was interested in such experimental work no matter how simple-minded the problem; in one discussion in Los Alamos on such questions, he asked to be given "interesting" numbers for computation of their continued fraction development. I named the quartic irrationality y given by the equations $y = 1/(x + y)$ where $x = 1/(1 + x)$ as one in whose development there might appear some curious regularities. Computations of many other numbers were planned, but it is not known to me whether this little project was ever pursued.

Game theory

This subject forms a new, rapidly developing chapter in present-day mathematical research; it is essentially a creation of von Neumann's. His fundamental work in this field has been described elsewhere, and I shall content myself with remarking that it presents some of his most fecund and influential work. It was Borel, in a note in the Comptes-Rendus in 1921, who first formulated a mathematical scheme describing strategies in a game between two players. The subject can, however, be dated as really originating in the paper of von Neumann. It is there that the fundamental "minimax" theorem is proved and the general scheme of a game between n players ($n \geq 2$) is formulated. Such schemata, quite apart from their interest and applications to actual games in economics, etc. introduced a wealth of novel combinatorial problems of purely mathematical interest. The theorem that Min Max = Max Min and the corollaries on the existence of saddle points of functions of many variables is contained in his 1937 paper. They are shown to be a consequence of a generalization of Brouwer's fixed-point theorem and of the following geometrical fact. Let S, T be two non-empty, convex, closed, and bounded sets contained in the Euclidean spaces R_n and R_m respectively. Let $S \times T$ be the direct product of these sets and V, W two closed subsets of it. Assume

197

that for every element x of S the set $Q(x)$ of all y such that (x, y) belongs to V is a closed convex and non-empty set. Analogously, for every y in T the set $P(y)$ of all x such that (x, y) belongs to W also has this property. Then the sets V and W have at least one point in common. This theorem, further discussed by Kakutani, Nash, Brown and others, plays a central role in the proofs of existence of "good strategies."

Economics

The now classical treatise by Oskar Morgenstern and John von Neumann, *Theory of games and economic behavior* contains an exposition of game theory in its purely mathematical form with a very detailed account of applications to actual games; and together with a discussion of some fundamental questions of economic theory introduces a different treatment of problems of economic behavior and certain aspects of sociology. The economist Oskar Morgenstern, a friend of von Neumann's in Princeton for many years, interested him in aspects of economic situations, specifically in problems of exchange of goods between two or more persons, in problems of monopoly, oligopoly and free competition. It was in a discussion of attempts to schematize mathematically such processes that the present shape of this theory began to take form.

The present numerous applications to "operational research," problems of communications and the statistical estimation theory of A. Wald either stem from or are drawing upon the ideas proposed and worked out in this monograph. We cannot outline in this chapter even the scope of these investigations. The interested reader may find an account of it in, e.g., L. Hurwicz's *The theory of economic behavior* and J. Marshak's *Neumann's and Morgenstern's new approach to static economics*.

Dynamics, mechanics of continua, meteorological calculations

In two papers written jointly with S. Chandrasekhar the following problem is considered. A random distribution of mass centers is assumed; these might be, for example, stars in a cluster or a cluster of nebulae. These masses are mutually attracting and in motion. The problem is to develop the statistics of the fluctuating gravitational field and the study of the motions of individual masses subject to the changing influence of the varying local distributions. In the first paper, the problem of the rate of the fluctuations in the distribution function for the force is solved through ingenious calculations, and a general formula is obtained for the probability distributions $W(F, f)$ of a gravitational field strength F and an associated rate of change f which is the derivative of F with respect to time. Among the results obtained is the theorem that for weak fields the probability of a change occurring in the field acting at a given instant of time is independent of the direction and magnitude of the initial field, while for strong fields, the probability of a change occurring in the direction of the initial field is twice as great as in a direction at right angles to it.

The second paper is devoted to a statistical analysis of the speed of fluctuations in the force per unit mass acting on a star which moves with a velocity V with respect to the centroid of the nearby stars. This problem is solved on the assumption of a uniform Poisson distribution of the stars and a spherical distribution of the local velocities. It is solved for a general distribution of different masses. An expression is derived for the correlations in the force acting at two very close points. The method gives the asymptotic behavior of the space correlations. Von Neumann was long interested in the phenomenon of turbulence. The writer remembers discussions in 1937 on the possibility of a statistical treatment of the Navier-Stokes equations by an analysis of hydrodynamical problems through replacement of the partial differential equations by a system of infinitely many total differential equations satisfied by the Fourier coefficients in the development of the Lagrangian functions in a Fourier series. A mimeographed report written by von Neumann for the Office of Naval Research in 1949, *Recent theories of turbulence*, constitutes a penetrating and lucid presentation of the ideas of Onsager and Kolmogoroff, and of other work up to that time.

With the beginning of the second World War, von Neumann undertook a study of problems presented by the motions of compressible gases and especially the perplexing phenomena of formation of discontinuities, e.g., shocks.

The greater part of his voluminous study in this field was prompted by problems arising in defense work. They were published in the form of reports.

It is impossible to summarize here this varied work; most of it is characterized by his incisive analytical technique and the customary clarity of logic. In the theory of interaction of colliding shocks, his contributions are especially noteworthy. One result is the first rigorous justification of the Chapman-Jouguet hypothesis concerning the process of detonation, that is, a combustion process initiated by a shock.

The formidable mathematical problems presented by the hydrodynamical equations of the motions of the earth's atmosphere fascinated von Neumann for a considerable time. With the advent of computing machines, a detailed numerical study at least of simplified versions of the problems became possible, and a large program of such work was started by him. At the Institute in Princeton, a meteorological research group was established, the plan was to attack the problem of numerical weather solution by a step-by-step investigation of models which were to approximate more and more closely the real properties of the atmosphere. A numerical investigation of truly 3-dimensional motions was impractical even on the most advanced electronic computing machines of the time.

The first highly schematized computations which von Neumann initiated dealt with a 2-dimensional model and for the most part in the so-called geostrophic approximation. Later, what might be called "2 + 1/2" dimensional hydrodynamical computations were performed by assuming two or three 2-dimensional models corresponding to different altitudes or pressure levels interacting with each other. This problem was dear to his mind, both because of its intrinsic mathematical interest, and because of the enormous technological consequences which a successful solution could have. He believed that our knowledge of the dynamics of controlling processes in the atmosphere, together with the development of computing machines, was approaching a level that would permit weather prediction. Beyond that, he believed that one could understand, calculate, and perhaps put into effect processes ultimately permitting control and change of the climate.

Theory and practice of computing on electronic machines, Monte Carlo method

Von Neumann's interest in numerical work had different sources. One stemmed from his original work on the role of formalism in mathematical logic and set theory, and his youthful work was concerned extensively with Hilbert's program of considering mathematics as a finite game. Another equally strong motivation came from his work in problems of mathematical physics including the purely theoretical work on ergodic theory in classical physics and his contributions to quantum theory. A growing exposure to the more practical problems encountered in hydrodynamics and in the manifold problems of mechanics of continua arising in the technology of nuclear energy led directly to problems of computation.

We have discussed briefly his interest in the problems of turbulence, general dynamics of continua, and meteorological calculations.

I remember quite well how, very early in the Los Alamos Project, it became obvious that analytical work alone was often not sufficient to provide even qualitative answers. The numerical work by hand and even the use of desk computing machines required a prohibitively long time for these problems. This situation provided the final spur for von Neumann to engage himself energetically in the work on methods of computation utilizing the electronic machines.

For several years von Neumann had felt that in many problems of hydrodynamics — in the propagation and behavior of shocks, and generally in cases where the non-linear partial differential equations describing the phenomena had to be applied in instances involving large displacements (that is in cases where linearization would not adequately approximate the true description) numerical work was necessary to provide heuristic material for a future theory.

This final necessity compelled him to examine, *from its foundations,* the problem of computing on electronic machines and, during 1944 and 1945, he formulated the now fundamental methods of translating a set of mathematical procedures into a language of instructions for a computing machine. The electronic machines of that time (e.g., the Eniac) lacked the flexibility and generality which they now possess in the handling of mathematical problems. Broadly speaking, each problem required a special and different system of wiring in order to enable the machine to perform the prescribed operations in a given sequence. Von Neumann's great contribution was the idea of a fixed and rather universal set of connections or circuits in the machine, a "flow diagram," and a "code" to enable a fixed set of connections in the machine to have the means of solving a very great variety of problems. While, *a priori* at least, the possibility of such an arrangement might be obvious to mathematical logicians, the execution and practice of such a universal method was far from obvious with the then existing electronic technology.

It is easy to underestimate, now, years after the inception of such methods, the great possibilities opened through such theoretical experimentation in problems of mathematical physics. The field is still new and it seems risky to make prophesies, but the already accumulated mass of theoretical experiments in hydrodynamics, magneto-hydrodynamics, and quantum-theoretical calculations, etc., allow one to hope that good syntheses may arise from these computations.

The engineering of the computing machines owes a great deal

to von Neumann. The logical schemata of the machines, the planning of the relative roles of their memory, their speed, the selection of fundamental "orders" and their circuits in the present machines bear heavily the imprint of his ideas. Von Neumann himself supervised the construction of a machine at the Institute for Advanced Study in Princeton, so as to acquaint himself with the engineering problems involved and at the same time to have at hand this tool for novel experimentation. Even before the machine was finished, which took longer than anticipated, he was involved in setting up and executing enormous computations arising in certain problems at the Los Alamos Laboratory.

One of these, the problem of following the course of a thermonuclear reaction, involved more than a billion elementary arithmetical operations and elementary logical orders. The problem was to find a "yes" or "no" answer to the question of propagation of a reaction. One was not concerned with providing the final data with great accuracy but, in order to obtain an answer to the original question, all the intermediate and detailed computations seemed necessary. It is true that guessing the behavior of certain elements of the problem, together with hand calculations, could indeed throw considerable light on the final answer, but in order to increase the degree of confidence in estimates thus obtained by intuition, an enormous amount of computational work had to be undertaken. This seems to be rather common in some new problems of mathematical physics and of modern technology. Astronomical accuracy is not required in the description of the phenomena; in some cases, one would be satisfied with predicting the behavior "up to 10 percent" and yet during the course of the calculations, the individual steps have to be kept as accurate as possible. The enormous number of elementary steps then poses the problem of estimating the reliability of final results and problems on the intrinsic stability of mathematical methods and their computational execution.

In receiving the Fermi prize of the Atomic Energy Commission, von Neumann was cited especially for his contribution to the development of computing on the electronic machines, so useful in many aspects of nuclear science and technology.

The electronic computing machines with their speed of computation surpassing that of the hand calculations by a factor of many thousands invite the invention of entirely new methods not only in numerical analysis in the classical sense, but in the very foundations of procedures of mathematical analysis itself. Nobody was more aware of these implications than von Neumann. A small example of what we mean here can be illustrated by the so-called Monte Carlo method. The methods of numerical analysis as developed in the past for hand work, or even for the relay machines, are not necessarily optimal for computations on the electronic machines. So, for example, it is obvious that instead of employing tables of elementary functions, it is more economical to compute the desired values directly. Next, it is clear that the procedures of integration of equations by reduction to quadratures, etc., can now be circumvented by schemes so complicated arithmetically that they could not even be considered for hand work, but are very feasible on the new machines. Literally dozens of computational tricks, "subroutines," e.g., for calculating elementary algebraical or transcendental functions, for solving of auxiliary equations, etc. were produced by von Neumann during the years following World War II. Some of this work, by the way, is not as yet generally available to the mathematical public, but is more widely known among the now numerous technological and scientific groups utilizing the computing machines in industrial or government projects. This work includes methods for finding eigenvalues and inversion of matrices, methods for economical search for extrema of functions of several variables, production of random digits, etc. Much of this exhibits the typical combinatorial dexterity, in some cases, of virtuoso quality, of his early work in mathematical logic and algebraical studies in operator theory.

The simplicity of mathematical formulation of the principles of mathematical physics hoped for in the nineteenth century seems to be conspicuously absent in modern theories. A perplexing variety and wealth of structure found in what one considered as elementary particles, seem to postpone the hopes for an early mathematical synthesis. In applied physics and in technology one is forced to deal with situations which, mathematically, present mixtures of different systems: For example, in addition to a system of particles whose behavior is governed by equations of mechanics, there are interacting electrical fields, described by partial differential equations; or, in the study of behavior of neutron-producing assemblies, one has, in addition to a system of neutrons, the hydrodynamical and thermodynamical properties of the whole system interacting with the discrete assembly of these particles.

From the point of view of combinatorics alone, not to mention the difficulties of analysis in the handling of several partial differential and integral equations, it is clear that at present, there is very little hope of finding solutions in a closed form. In order to find, *even only qualitatively*, the properties of such systems, one is forced to look for pragmatic methods.

We decided to look for ways to find, as it were, homomorphic images of the given physical problem in a mathematical schema which could be represented by a system of fictitious "particles" treated by an electronic computer. It is especially in problems involving functions of a considerable number of independent variables that such procedures would be applied. To give a very simple concrete example of such a Monte Carlo approach, let us consider the question of evaluating the volume of a subregion of a given n-dimensional "cube" described by a set of inequalities. Instead of the usual method of approximating the volume required by a systematic subdivision of the space into its lattice points one could select, *at random*, with uniform probability, a number of points in space and determine (on the machine) how many of these points belong to the given region. This proportion will give us, according to elementary facts of probability theory, an approximate value of the relative volumes, *with the probability as close to one* as we wish, by employing a sufficient number of sample points.

As a somewhat more complicated example, consider the problem of diffusion in a region of space bounded by surfaces which partly reflect and partly absorb the diffusing particles. If the geometry of the region is complicated, it might be more economical to try to perform "physically" a large number of such random walks rather than to try to solve the integro-differential equations classically. These "walks" can be performed conveniently on machines and such a procedure in fact reverses the treatment which in probability theory reduces the study of random walks to the study of differential equations.

Another instance of such methodology is, given a set of functional equations, to attempt to transform it into an equivalent one which would admit of a probabilistic or game theory interpretation. This latter would allow one to play, on a machine, the game illustrating the random processes and the distributions obtained would give a fair idea of the solution of the original equations. Better still, the hope would be to obtain directly a "homomorphic image" of the behavior of the physical system in question. It has to be stated that in many physical problems presently considered, the differential equations originally obtained by certain idealizations, are not, so to say, very sacrosanct any more. A direct study of models of the system on computing machines may possess a heuristic value, at least.

A great number of problems were treated in this fashion towards the end of the war and in following years by von Neumann and the writer. At first, the probabilistic interpretation was immediately suggested by the physical situation itself. Later, problems of the third class mentioned above were studied. A theory of such mathematical models is still very incomplete. In particular, estimates of fluctuations and accuracy are not as yet developed. Here again, von Neumann contributed a large number of ingenious ways, for example by playing suitable games, of producing sequences of numbers in the given probability distributions. He also devised probabilistic models for treatment of the Boltzmann equation and important stochastic models for some strictly deterministic problems in hydrodynamics. Much of this work is scattered throughout various laboratory reports or is still in manuscript. One certainly hopes that in the future, an organized selection will be available to the mathematical public.

Theory of automata, probabilistic logic

This work, like that in game theory, has stimulated, since von Neumann's death, a wide and increasingly expanding number of studies and seems to me to rank with his most fertile ideas. Here a combination of his interest in mathematical logic, computing machines, mathematical analysis, and the knowledge of problems of mathematical physics, come to bear fruit in new constructions. The ideas of Turing, McCulloch, and Pitts on the representation of logical propositions by electrical networks or idealized nervous systems inspired him to propose and outline a general theory of automata. Its notions and terminology come from several fields — mathematics, electrical engineering, and neurology. Such studies now promise more conquests of mathematics in its ability to formalize, perhaps at first on an extremely simplified level, the workings of an organism and of the nervous system itself.

Nuclear energy, work at Los Alamos

The discovery of the phenomenon of fission in uranium caused by absorption of neutrons with a consequent release of more neutrons came just before the outbreak of the Second World War. A number of physicists realized at once the possibility of a vast release of energy in an exponential reaction in a mass of uranium, and discussions started on quantitative evaluation of arrangements which would lead to utilization of this new source of energy.

Theoretical physicists form a much smaller and more closely knit group than mathematicians and, in general, the interchange of results and ideas is more rapid among them. Von Neumann, whose work in foundations of quantum theory brought him early into contact with most of the leading physicists, was aware of the new experimental facts and participated, from the beginning, in their speculations on the enormous technological possibilities latent in the phenomena of fission. The outbreak of war found him already engaged in scientific work connected with problems of defense. It was not until late in 1943, however, that he was asked by Oppenheimer to visit the Los Alamos Laboratory as a consultant and began to participate in the work which was to culminate in the construction of the atomic bomb.

As is now well known, the first self-sustaining nuclear chain reaction was established by a group of physicists headed by Fermi in Chicago on December 2, 1942, through the construction of a pile, an arrangement of uranium and a moderating substance where the neutrons are slowed down in order to increase their probability of causing further fissions. A pile forms a very large object and the time for the *e*-folding of the number of neutrons is relatively long.

The project established at Los Alamos had as its aim to produce a very fast reaction in a relatively small amount of the 235 isotope of uranium or plutonium, leading to an explosive release of a vast amount of energy. The scientific group began to assemble in late spring of 1943 and by fall of that year a great number of eminent theoretical and experimental physicists were settled there.

When von Neumann arrived in Los Alamos, diverse methods of assembling a critical mass of fissionable material were being examined; no scheme was a priori certain of success, one of the problems being whether a sufficiently fast assembly is possible before the nuclear reaction would lead to a mild or mediocre explosion preventing the utilization of most of the material.

E. Teller remembers how Johnny arrived in Lamy (the railroad station nearest Los Alamos), was brought up to the "hill," surrounded at that time by great secrecy, in an official car:

"When he arrived, the Coordinating Council was just in session. Our Director, Oppenheimer, was reporting on the Ottawa meeting in Canada. His speech contained lots of references to most important people and equally important decisions, one of which affected us closely: We could expect the arrival of the British contingent in the near future. After he finished the speech he asked whether there were any questions or comments. The audience was impressed and no questions were asked. Then Oppenheimer suggested that there might be questions on some other topics. After a second or two a deep voice (whose source has been lost to history) spoke, 'When shall we have a shoemaker on the Hill?' Even though no scientific prob-

lem was discussed with Johnny at that time, he asserted that as of that moment he was fully familiar with the nature of Los Alamos."

The atmosphere of work was extremely intense at that time and more characteristic of university seminars than technological or engineering laboratories by its informality and the exploratory and, one might say, abstract character of the scientific discussions. I remember rather vividly that it was with some astonishment that I found, upon arriving at Los Alamos, a milieu reminiscent of a group of mathematicians discussing their abstract speculations rather than of engineers working on a well defined practical project — discussions were going on informally often until late at night. Scientifically, a striking feature of the situation was the diversity of problems, each equally important for the success of the project. For example, there was the problem of the distribution, in space and time, of the neutrons whose number increases exponentially; equally important were the problems of following the increasing deposition of energy by fissions in the material of the bomb, the calculation of hydrodynamical motions in the explosion, the distribution of energy in the form of radiation; and finally, following the course of the motions of the material surrounding the bomb after it has lost its criticality. It was vital to understand all these questions which involved very different mathematical problems.

It is impossible to detail here the contributions of von Neumann; I shall try to indicate some of the more important ones. Early in 1944 a method of *implosion* was considered for the assembly of the fissionable material. This involves a spherical impulse given to the material, followed by the compression. Von Neumann, Bethe, and Teller were the first to recognize the advantage of this scheme. Teller told him about the experimental work of Neddermeyer and collaborated with Von Neumann on working out the essential consequences of such spherical geometry. Von Neumann came to the conclusion that one could produce exceedingly great pressures by this method and it became clear in the discussion that great pressures would bring about considerable compressions as well. In order to start the implosion in a sufficiently symmetrical manner, the original push given by high explosives had to be delivered by simultaneously detonating it from many points. Tuck and von Neumann suggested that it be supplemented by the use of high explosive lenses.

We mentioned before von Neumann's ability, perhaps somewhat rare among mathematicians, to commune with the physicists, understand their language, and to transform it almost instantly into a mathematician's schemes and expressions. Then, after following the problems as such, he could translate them back into expressions in common use among physicists.

The first attempts to calculate the motions resulting from an implosion were extremely schematic. The equations of state of the materials involved were only imperfectly known, but even with crude mathematical approximations one was led to equations whose solution was beyond the scope of explicit analytical methods. It became obvious that extensive and tedious numerical work was necessary in order to obtain quantitatively correct results and it is in this connection that computing machines appeared as a necessary aid.

A still more complicated problem is that of the calculation of the characteristics of the nuclear explosion. The amount of energy liberated in it depends on the history of the outward motions which are, of course, governed by the rate of energy deposition and by the thermodynamic properties of the material and radiation at the very high temperatures which are generated. One had to be satisfied for the first experiment with approximate calculations; however, as mentioned before, even the order of magnitude is not easy to estimate without intricate computations. After the end of the war the desire to economize on the material and to maximize its utilization prompted the need for much more precise calculations. Here again von Neumann's contributions to the mathematical treatment of the resulting physical questions were considerable.

Already during the war, the possibilities of *thermonuclear* reactions were considered, at first only in discussions, then in preliminary calculations. Von Neumann participated actively as a member of an imaginative group which considered various schemes for making possible such reactions on a large scale. The problems involved in treating the conditions necessary for such a reaction and in following its course are even more complex mathematically than those attending a fission explosion (whose characteristics are indeed a prerequisite for following the larger problem). After one discussion in which we outlined the course of such a calculation, von Neumann turned to me and said, "Probably in its execution we shall have to perform more elementary arithmetical steps than the total of all the computations performed by the human race heretofore." We noticed, however, that the total number of multiplications made by the school children of the world in the course of a few years sensibly exceeded that of our problem!

Limitations of space make it impossible to give an account of the innumerable smaller technical contributions of von Neumann welcomed by physicists and engineers engaged in this project.

Von Neumann was very adept in performing dimensional estimates and algebraical and numerical computations in his head without using pencil and paper. This ability, perhaps somewhat akin to the talent of playing chess blindfolded, often impressed physicists. My impression was that von Neumann did not visualize the physical objects under consideration but rather treated their properties as logical consequences of the fundamental physical assumptions; but he was able to play a deductive game with these astonishingly well!

One trait of his scientific personality, which made him very much liked and sought after by those engaged in applications of mathematical techniques, was a willingness to listen attentively even to questions sometimes without much scientific import, but presenting the combinatorial attractions of a puzzle. Many of his interlocutors were helped actively or else consoled by knowing that there is no magic in mathematics known to anyone containing easy answers to their problems. His unselfish willingness to be involved in perhaps too diverse and certainly too numerous activities where mathematical insight might be useful (they are so increasingly common in technology now-adays) put severe demands on his time. In the years following the end of the Second World War, he found himself torn between conflicting demands on his time almost every moment.

Von Neumann strongly believed that the technological revolution initiated by the release of nuclear energy would cause more profound changes in human society, in particular in the development of science, than any technological discovery made in the previous history of the race. In one of the very few instances of talking about his own lucky guesses, he told me that, as a very young man, he believed that nuclear energy would be made available and change the order of human activities during his lifetime!

He participated actively in the early speculations and deliberations on the possibility of *controlled* thermonuclear reactions. When in 1954 he became a member of the Atomic Energy Commission, he worked on the technical and economical problems relating to the building and operation of fission reactors. In this position he also spent a great deal of time in the organization of studies of mathematical computing machines and the means to make them available to universities and other research centers.

This fragmentary account of von Neumann's diverse achievements and this cursory peregrination through the mathematical disciplines in which he left so many permanent imprints, may raise the question whether there was a thread of continuity throughout his work.

As Poincaré phrased it: "*Il y a des problèmes qu'on se pose et des problèmes qui se posent.*" Now, some eighty years after the great French mathematician formulated this indefinite distinction, the state of mathematics presents this division in a more acute form. Many more of the objects considered by mathematicians are their own free creations, often, so to say, special generalizations of previous constructions. These are sometimes originally inspired by the schemata of physics, others evolve genetically from free mathematical creations — in some cases prophetically anticipating the actual patterns of physical relations. Von Neumann's thought was obviously influenced by both tendencies. It was his desire to preserve, so far as possible, the connection between the pyramiding mathematical constructions and the increasing combinatorial complexity presented by physics and the natural sciences in general, a connection which seems to be growing more and more elusive.

Some of the great mathematicians of the eighteenth century, in particular Euler, succeeded in incorporating into the domain of mathematical analysis descriptions of many natural phenomena. Von Neumann's work attempted to cast in a similar role the mathematics stemming from set theory and modern algebra. This is of course, nowadays, a vastly more difficult undertaking. The infinitesimal calculus and the subsequent growth of analysis through most of the nineteenth century led to hopes of not merely cataloguing, but of understanding the contents of the Pandora's box opened by the discoveries of physical sciences. Such hopes are now illusory, if only because the real number system of the Euclidean space can no longer claim, algebraically, or even only topologically, to be the unique or even the best mathematical substratum for physical theories. The physical ideas of the 19th century, dominated mathematically by differential and integral equations and the theory of analytic functions, have become inadequate. The new quantum theory requires on the analytic side a set-theoretically more general point of view, the primitive notions themselves involving probability distributions and infinite-dimensional function spaces. The algebraical counterpart to this involves a study of combinatorial and algebraic structures more general than those presented by real or complex numbers alone. Von Neumann's work came at a time when the whole complex of ideas stemming from Cantor's set theory and the algebraical work of Hilbert, Weyl, Noether, Artin, Brauer, and others could be exploited for this purpose.

Another major source from which general mathematical

investigations are beginning to develop is a new kind of combinatorial analysis stimulated by the recent fundamental researches in the biological sciences. Here, the lack of general method at the present time is even more noticeable. The problems are essentially non-linear, and of an extremely complex combinatorial character; it seems that many years of experimentation and heuristic studies will be necessary before one can hope to achieve the insight required for decisive syntheses. An awareness of this is what prompted von Neumann to devote so much of his work of his last ten years to the study and construction of computing machines and to formulate a preliminary outline for the study of automata.

Surveying von Neumann's work and seeing how ramified and extended it is, one could say with Hilbert: "One is led to ask oneself whether the science of mathematics will not end, as has been the case for a long time now for other sciences, in a subdivision of separate parts whose representatives will barely understand each other and whose connections will continue to diminish? I neither think so nor hope for this; the science of mathematics is an indivisible whole, an organism whose vital force has as its premise the indissolubility of its parts. Whatever the diversity of subjects of our science in its details, we are nonetheless struck by the equivalence of the logical procedures, the relation of ideas in the whole of science and the numerous analogies in its different domains . . . ." Von Neumann's work was a contribution to this ideal of the universality and organic unity of mathematics.

CHAPTER 17

VON NEUMANN: THE INTERACTION OF MATHEMATICS AND COMPUTING

THE TITLE OF THIS CHAPTER is: Von Neumann, the interaction of mathematics and computing, but it is very hard to separate sharply mathematics from physics in this connection. I'd go further and stress the very great possibilities that the same set of ideas, the same set of technological developments, can have in other natural sciences — primarily biology, for example, and that soon. One could call this the "music of the future," and I think this aspect of the future is something that should not be neglected, even in a chapter devoted to the past.

It must have been in 1938 that I first had discussions with von Neumann about problems in mathematical physics, and the first I remember were when he was very curious about the problem of mathematical treatment of turbulence in hydrodynamics. I think he discussed this with Norbert Wiener also shortly before. He was fascinated by the role of the Reynolds number, a dimensionless number, a pure number because it is the ratio of two forces, the inertial one and the viscous, and has the following importance: When its value surpasses a critical size, about 2000, the regular laminar flow, as it is called, becomes highly irregular and turbulent. Both Wiener in a general way and von Neumann, who I think knew more practical physics than Norbert, wanted to find an explanation or at least a way to understand this very puzzling large number. Small numbers like π and e, are of course very frequent in physics, but here is a number of the order of thousands and yet it is a pure number with no dimensions: it does whet our curiousity.

I remember that in our discussions von Neumann realized that the known analytical methods, the methods of mathematical analysis, even in their most advanced forms, were not powerful enough to give any hope of obtaining solutions in closed form. This was perhaps one of the origins of his desire to try to devise methods of very fast numerical computations, a more humble way of proceeding. Proceeding by "brute force" is considered by some to be more low-brow. I was not present at the discussions between Norbert Wiener and Johnny von Neumann — he just told me about them. A little later these two men had apparently developed different philosophies. Wiener thought that the computers (if you could use the word at that time) would be more in the nature of analogue machines than digital. Von Neumann maintained the opposite view. Wiener thought of the hormonal activity of the human brain — obviously there is no mechanical relay system in our brain — so he thought that the big developments would go in the direction of some kind of system of fluids, whereas von Neumann from the start was thinking of developing a digital or binary or purely discrete system.

I remember also discussions about the possibilities of calculating the weather at first only locally, and soon after that, about how to calculate the circulation of meteorological phenomena around the globe. There is of course already available from the last century some marvelous theoretical work, Laplace's and others' but it is not detailed enough to enable meteorologists to predict the motion of air locally, or on the largest scale round the globe. Some progress has been made since that time, but not nearly as much as in some other fields that I shall mention.

I can mention two cases in other fields of mathematical physics in which computers have played a decisive role and in which progress has been more obvious: the study of the equilibrium or even of the evolution of a star, and the calculations without which it would be impossible to predict the behavior of star clusters or gravitating masses of gas. In some ways the original intention or motivation is not what was immediately followed with rapid success. I think this is typical of many developments in technology. The original application of a new fact or tool is not what ultimately turns out to be its greatest application or achievement. In this connection I remember that when a release of nuclear energy was discovered in Chicago and developed in Los Alamos, with all its terrifying consequences, von Neumann remarked that the first use of naphtha or petroleum, as we now call it, was as a laxative. Look what happened in the following centuries! This is merely to point out that nuclear energy will have more interesting and certainly more beneficial effects and applications than its original one, the bombs.

The first papers that von Neumann wrote as a young man around 1924-1925 were in mathematical logic and the study of formal systems. It is perhaps a matter of chance, that computer development became possible only by a confluence of at least two entirely different streams. One is the purely theoretical study of formal systems. The study of how to formalize a description of natural phenomena or even of mathematical facts. The whole idea of proceeding by a given set of rules from a given set of axioms was studied successfully in this connection. The second stream is the technological development in electronics, which came at just the right time. Of course, the war greatly accelerated the availability of funds and effort just a few years later.

Remember that for many years von Neumann was very much a pure mathematician. It was only, to my knowledge, just before World War II that he became interested not only in mathematical physics but also in more concrete physical problems. His book on quantum theory is very abstract and is, so to say, about the grammar of the subject. Now it is extremely important to attempt even only tentatively to put a rigorous foundation to a new part of physics, and it is a valuable and important work. But it did not, it seems to me, contribute directly to any truly new insights or new experiments — there are probably some physicists who might dispute this point of view, but in the large it seems to me that it is so.

Already some years before the war von Neumann expected

a catastrophe. He thought there would come a great conflict involving also the United States. Living here and having come from Europe he was in a good position to see further. On the whole people who lived only in Europe or here in the States probably did not sense the currents as well as he did.

My own involvement with problems of a more practical physical nature started when I came to Los Alamos. I arrived during the early days of 1944 and learned right away what the project was working on, was introduced to a number of physicists, and, when I came into one of the offices, found to my surprise, von Neumann (who came frequently for periods of a week or two), Teller, who was with him, and some others. The blackboard was filled with very complicated equations that you could encounter in other forms in other offices. This sight scared me out of my wits: looking at these I felt that I should never be able to contribute even an epsilon to the solution of any of them. But during the following days, to my relief, I saw that the same equations remained on the blackboards. I noticed that one did not have to produce immediate solutions.

Even though I was not an applied mathematician, I knew some physics in a sort of platonic way, having always been interested in quantum theory and ideas of relativity and in astrophysics especially. Little as I already knew about partial differential equations or integral equations, I could feel at once that there was no hope of solution by analytical work that could yield practical answers to the problems that appeared.

One of the first problems that was crucial to the success of the whole project was the behavior of an implosion of a spherical system. The word itself was highly classified during the war. It would have been a terrible breach of security even to utter it, and this was true even for a couple of years after the war. But then it became a *"secret de polichinelle;"* everybody at least had heard of the word. The idea was to compress a mass of material into higher density by surrounding it with explosives and try to figure out what pressures and densities would be achievable, and how the material could get to such configurations. It was not enough to know the answers within a factor of two or three. One had to have a more precise numerical value, an estimate of the pressure, say, within 10%. This was really impossible to guess, or to derive from an analysis of theoretical dimensional reasonings alone. We had a long discussion about using a purely "brute force" approach.

Being so erudite in many fields of mathematics in addition to his own, von Neumann tried to work out with some of his collaborators at least part way analytical methods, to find out what would happen when the material was pushed together. However, the accuracy or reliability was quite unsatisfactory. I was trying to press him to try at least some step-by-step numerical procedures assuming, of course, knowledge of the equations of state and so on — these were known with some accuracy. Von Neumann was at that time a consultant to the Aberdeen Ballistics Laboratory, and he knew the computing machines there. I did convince him; we received administrative support for getting all possible means to enable one to calculate implosions more exactly. He did a lot of this work around 1948–1949 on the ENIAC at Aberdeen.

The study of the implosion problem gave one of the great impulses to the development of fast computers. There were many others of the same sort of equal importance to Los Alamos, such as the equations of state themselves. And because it had to solve these problems Los Alamos, consciously or not, made a great and fundamental contribution to the development of computing.

So far I have been speaking of mathematical physics, in which such great developments took place. But in mathematics itself these were slower. It was more of a luxury, since machine time was expensive, to try to compute things of only pure mathematical interest. The calculations and experiments were at first in the nature of fillers: There was sometimes a free period of time on the machine, and one could amuse oneself by trying to work on problems in pure mathematics. So the interaction of computers and mathematics started almost playfully. But remember the theory of probability, which is now so fundamental in many areas of theoretical physics, statistical mechanics, quantum theory, etc. and pervades all kinds of very theoretical and of course also more mundane kinds of mathematics. The natural material to play with in those early days was in combinatorial problems and in number theory. As far as number theory goes, the use of computers has an older history. I was still a student in Poland in the late 1920s when I first heard from my professor Hugo Steinhaus of a mathematician in California who had devised a mechanical way to find primes and to study some of their properties using a system of cylinders with holes. That was D. N. Lehmer, the father; his son, Professor D. H. Lehmer, has worked with this machine and is one of the pioneers of the use of electronic computers in number theory.

Of course, when it comes to operations in prime numbers or other questions in number theory, and to the consideration of very large numbers, this becomes possible only with very fast machines. From the outset the most impressive thing about electronic computers has been their speed. One of the fears expressed by mathematicians about the use of computers was that the interesting numbers in combinatorial mathematics are so large, even in relatively simple problems, that even the biggest computers could not begin to touch the general cases so as to give us confidence in the asymptotic behavior. This is less and less true. There are now examples of situations in which many billions of alternatives have been sieved through in a few hours. One such example is the famous four-color problem: if you have a map on a plane or a sphere and want to color the countries so that countries with a common boundary always have different colors, how many colors do you need? The conjecture is that this can always be done with four colors, but nobody was able to prove this until very recently. Strangely enough this problem can be studied on a computer. I studied it in a mutilated form and a former student of mine in Colorado managed to prove the conjecture for an infinite band seven countries wide — for which millions of cases had to be examined. Recently the full problem has been solved, by a method that would not have been possible without a fast computer.

The reason I mention all this is that mathematicians, who tend perhaps to be a little snobbish, are not satisfied with the purely finite: they need at least an inkling of infinity! So in order to "hook" mathematicians in my own modest attempts to interest them in some problems of mathematical biology, I always try to formulate the problems in such a way that they have a sense not only for finite assemblies but for cases with true infinity. This is merely a sort of psychological stratagem, but it will cease to be so in the future when some "elements" of true infinity may be, by use of so-called quantifiers in algebra, *mirrored* in operations on finite computers.

When I talked to von Neumann about this I learned that it was also his hope and belief. I use the word quantifier; perhaps I should explain its meaning.

The machines can express well the Boolean — Aristotelian expression in logic, which consists of the words *and* or *not*. Mathematicians however, greatly love expressions like "there exists an X such that . . . ." This "exists" is a quantifier. The other quantifier is "for all," as in "for all X such and such is true." There are just these two quantifiers, which seems so very innocent, but they have a character different from the purely Aristotelian operations. Somehow they have not yet been incorporated, even by approximation, on the computers, except in the most primitive and too literally finite way. So this is one of the possibilities for the future, beyond the merely continuously improving way of surveying larger and larger numbers of cases, which ultimately might give insights into the true nature of physical laws by natural induction.

It seems that the great advances we now see, great as they are interesting, are still only in their infancy. I feel that the future holds much more. The present machines can execute only instructions given in advance, and as far as the logical operations or the arithmetical operations go, they are limited to Boolean operations, Aristotelian logic, and the four arithmetic operations or else ensuing evaluations of integrals, derivatives, etc. There is no doubt that a more general abstract way of following and studying the development of mathematical symbols will be rendered possible to a much larger extent by new machines.

I should like to propagandize, if that is needed, some work to develop the use of machines operating in parallel, on many channels at the same time. Up to now our computers essentially follow the course of making one step, one deduction at a time, this of course with some reservations and caveats. I hope that in the future, perhaps even the near future, there will be machines built to imitate more closely certain features of the brain and of the nervous system, which certainly work simultaneously on very many channels.

After the war, von Neumann was interested — starting with the analogies between the computer and some of the mental processes — in the mysteries of the workings of the nervous system and the brain itself. He published several papers on this subject. We are not yet at the stage where one can say anything very meaningful even only about the physiological nature of human memory. There are fundamental controversies among physiologists: Does the memory reside in molecules or is it maintained as a system of currents between the neurons of the brain? That simple dichotomy is not resolved. Some partial understanding of how the memory works will probably come in a not too distant future, and with it more effective means of creating work in parallel on computers that will be so superior to even the best we have now. Von Neumann realized that the search mechanism used by the brain must be very different from the ones we use on our computers, and this, when we understand it, must be the one we shall then use.

Our conscious reasoning, the things we write down, appears to be linear. But the real search in our memory and the process of thought certainly proceed simultaneously on very many channels. I remember in discussions with von Neumann the great marvel was that there are many billions of neurons in the brain and, as he told me at that time, there were perhaps as many as 50 or 100 connections between some of them. Today this number turns out to be perhaps several thousand instead of 50; in the central region it may be 100,000. So complications grow not only in the foundations of physics or in astrophysics but even in anatomy. Everything in science seems to become much more complicated than we once thought.

CHAPTER 18

JOHN VON NEUMANN
ON COMPUTERS AND THE BRAIN

THE BOOK "THE COMPUTER AND THE BRAIN" was published more than a year after von Neumann's death. It is hardly more than an introduction to a monograph on a subject which preoccupied him during the last years of his life. It has been prepared from an incomplete manuscript of the Silliman Lectures which he was to give at Yale University in 1956. As Mrs. von Neumann explains in the introduction, the lectures were written during his fatal illness. In places the presentation of the material lacks his characteristic style. Nevertheless the book, like everything von Neumann wrote, remains highly original and intensely stimulating.

Von Neumann became interested in the possibilities of electronic computing machines during the Second World War. In the beginning he was primarily concerned with the logic of the operation of such machines, but he was the first to devise a means by which a machine with fixed circuits could deal flexibly with a variety of mathematical problems. Before he had entered the field, the solution of each problem required a different set of wiring connections.

It was the problems of mathematical physics that arose at the Los Alamos Scientific Laboratory which created the need to plan massive computations on the new machines. The solution of one of these problems may require billions of elementary steps: additions, multiplications, and so on. Beyond the mere bulk

of repetitive computations the problem may involve combinations and logical operations of considerable complexity. The
program of calculation must be prepared in advance, first in
a general form. This is done by means of a flow diagram, which
is then elaborated by a detailed set of instructions — a "code."
This very scheme of a flow diagram and a code is due to von
Neumann, along with many other concepts now commonplace
in the art of computing.

It was von Neumann's belief that computing machines would
not only help solve existing problems but also open new perspec-
tives in mathematics and physics. Obviously the machines could
be used to test tentative theories by computations too laborious
and too lengthy for manual methods. But, more important, they
could provide the imaginative investigator with new "experimental" material which could suggest new theories.

From the beginning von Neumann was intrigued by the
similarities and differences between the operation of computing
machines and the working of the human brain. He envisaged
a general theory of automata, and even of organisms, but his
ideas remained undeveloped. The present book is an approach
to an understanding of the nervous system from the point of
view of the mathematician. Von Neumann stated that, since
he was not a nerve physiologist or a psychologist, he would
restrict himself to the logical and statistical features of the
elements of the brain. He felt that a mathematical study of the
nervous system would not only lead to a better understanding
of this system but also could affect the future development of
mathematics itself.

The book is divided into two parts: (1) the computer and (2)
the nervous system. The first describes in terms comprehensible
to the layman (but with insufficient detail) what the characteristics
of a computing machine are, how it works and what it can do.
The discussion concerns both digital and analogue computers.
Present machines have organs to perform the basic arithmetical
operations: addition, subtraction, multiplication and division.
In the digital computer these operations are performed on
numbers stored as a sequence of markers. In the analogue

machine numbers are represented not by symbols but by physical quantities, for example, an electric current. The strength of the current can be added and subtracted; multiplication and division are more difficult, but can be accomplished by special devices. A mathematical machine also needs a memory in which numbers generated by one operation can be stored until they are needed for another.

The second part of the book deals with the nervous system in much the same language. The assumption has been made that the operation of the nervous system is digital, a view that is strongly suggested by the facts of nerve physiology. The basic unit of the nervous system is the nerve cell, or neuron, which generates and propagates nerve impulses. The nature of this impulse is not simple; in some respects the impulse is electrical, and in others it may be considered chemical and mechanical. The distinction between these terms may be clear in the macroscopic world, but on the scale of molecules they tend to merge.

A nerve impulse travels along the axon which branches out from the neuron. Associated with the impulse is a potential of about 50 millivolts, which lasts for about a thousandth of a second. During this disturbance there are chemical changes in the axon, and probably mechanical changes also. All these changes are reversible; after the impulse has passed, the axon reverts to its original state.

The digital character of the nervous system has the basis that the neuron may fire a nerve impulse or it may not. Thus the operation of the nervous system corresponds to the binary system of a digital computer, in which there are only two basic symbols: 0 and 1 ("yes" and "no"). It takes a neuron somewhere between a hundredth and a ten-thousandth of a second to react to a stimulus. The reaction time of a transistor, which may be a component of a digital computer, is between a billionth and a ten-billionth of a second. With respect to speed, then, the artificial components are ahead by a factor of 10 million or 100 million.

When it comes to the size and number of components the situation is entirely different. The size of a neuron varies widely from cell to cell. However, to compare the size of the logically active part of a nerve cell with that of the logically active part of a transistor we can use rough linear dimensions. The thickness of the neuron's outer membrane, in which most of the changes associated with the nerve impulse occur, is of the order of a ten thousandth of a centimeter. The corresponding quantity in the vacuum tube, the distance between grid and cathode, is between a tenth and a hundredth of an inch. In the transistor the distance between the "emitter" and the "control electrode" is less than a hundredth of a millimeter. Thus in linear dimensions the natural components are more compact by a factor of the order of 10.

The volume occupied by the human brain is of the order of 1,000 cubic centimeters; the number of neurons is estimated at 10 billion or more. This allows a space of about a ten-millionth of a cubic centimeter for each neuron. Today a few thousand vacuum tubes would certainly occupy tens of cubic feet. By replacing vacuum tubes with transistors this space is reduced by a factor of 100 or so. But the economy of packing in the natural system still leads that in the artificial ones by a factor like 100 million or a billion.

The energy consumption of natural and artificial components can also be compared. The human brain dissipates energy at the rate of about 10 watts: a billionth of a watt per neuron. A vacuum tube dissipates about 50 watts; a transistor, perhaps a tenth of a watt. Here again the natural components lead the artificial by a factor of the order of 100 million or a billion.

So nature seems to favor a system with relatively slow components but a great many of them. Electronic computers utilize fast components, but relatively few of them. This suggests that the human nervous system handles many items of information simultaneously, whereas modern computing machines tend to do one thing at a time, or relatively few things at a time. In other words, the function of the nervous system is largely "parallel" and that of computers is essentially "serial." This very important point is brought out in von Neumann's discussion. It should be noted that, in general, parallel and serial operations cannot simply be substituted for each other. Not all serial operations can immediately be converted into parallel ones. For example, a step in a sequence of calculations performed by a computer may depend on the result of other steps. Conversely, a parallel procedure generally cannot be adapted to the serial processes of a computer without creating new requirements for the computer, specifically in its memory.

Von Neumann now discusses the functions of the nervous system which go beyond the simple digital picture. He introduces a mathematical scheme which distinguishes between different kinds of neurons. Some neurons, for example, respond to external stimuli such as light and sound; others, called internal receptors, respond only to nerve impulses. Some light receptors respond not to the level of illumination but to changes in it. Some neurons are activated by the stimulation of certain of their branches, and not by the stimulation of others. All this suggests that the nervous system functions not only like a digital computer but also like an analogue machine. A realistic mathematical description of this versatile system has yet to be written.

So far we have considered organs which correspond to the arithmetical components of a computer. What about memory? At the outset of his discussion von Neumann states that we are as ignorant of the nature and location of the human memory as were the Greeks, who placed the mind in the diaphragm. We only know that the capacity of this memory is remarkably large. A memory contains a certain maximum amount of information, which in the usage of modern information theory can be converted into binary digits ("bits"). As an example, a computer memory which can hold 1,000 eight-digit decimal numbers has a capacity of 26,640 bits. The memory of a reasonable electronic computer has a capacity of 100,000 to 1,000,000 bits. Now let us assume, first, that in the human memory nothing is ever truly erased (there is some evidence for this assumption); second, that a receptor in the nervous system receives about 10 distinct digital impressions a second; third (as before), that there are 10^{10} neurons; and, fourth, that the life span is 60 years, or 2×10^9 seconds. Thus we have 10×10^{10} times 2×10^9, or 2×10^{20} bits. Compared to the 10^5- or 10^6-bit memory of a computer, this is an impressive number.

Modern ideas on the nature of memory are still vague and unsatisfactory. One idea is that the stimulation of a nerve cell slightly changes the requirements for further stimulation; the memory would reside in the change of these conditions. Another proposed scheme is that connections among nerve cells change with stimulation; here the memory would reside in the change of connections. Von Neumann speaks of the genetic mechanism as a memory. It could be argued that certain chemical systems of the body which perpetuate themselves represent memory elements. This is reminiscent to those who are familiar with electronic computers, because a memory element in such a machine automatically regenerates its state. Von Neumann argues that some systems of nerve cells which act in a cyclic or periodic manner constitute a memory. This reminds one of computer "flip-flops," in which four transistors control each other.

After this discussion von Neumann returns to the interplay of digital and analogue mechanisms in living organisms. Here the genes provide a cogent example. The genes themselves are part of a digital system, that is to say, they bear a code which determines the characteristics of the organism. The effect of the genes, however, is to produce biological catalysts (enzymes), whose role is rather like that of analogue components in a computer.

The book also takes up the concept of the "complete" code. This is a set of instructions which describes each and every step in the solution of a problem by a machine. It can be contrasted to the more general idea of the "short" code, which assumes a machine with capabilities beyond those of present computers. The short code is a brief set of instructions for a machine which will elaborate and complete these instructions by itself. The concept of the short code has already been applied, in a limited way, to actual computers. It is possible to present a machine with a "verbal" description of a problem; the machine will then write the complete code by itself and follow it.

Ultimately, of course, both machine and organism must have

a complete code. The question now arises: How precisely are these instructions supposed to be carried out? In a computing machine they must be carried out very precisely, or the errors will quickly add up. Von Neumann believed, however, that in the nervous system the notation is not exclusively digital but is to some extent statistical. He discusses this point with reference to his earlier work on how to build reliable machines from nonreliable components.

Von Neumann's final paragraphs are devoted to the speculation that there are two kinds of communication in the human brain. The first involves logic and is akin to mathematics as we now know it; the second is a kind of general language. This language, which may correspond to the short code, must be quite different from languages in the usual sense. It may, for example, operate with analogies.

This little volume will leave the reader with a sharper awareness of the loss represented by von Neumann's premature death. It was not given to him to pursue these speculations and render them into mathematical form. But his ideas will have great value to further investigations.

We are only beginning to realize the potentialities of the large electronic computer. In the opinion of this reviewer the next steps in its evolution will involve not only greater speed and larger memory capacity but also a more intimate connection between the machine and its user. The usefulness of the machine would be greatly increased if the results of its computations could be quickly displayed and if its orders could be quickly changed. If the machine can be made to operate more like the human nervous system, it may stimulate the imagination of its user to a greater degree. Certainly it seems that future machines will perform many operations in parallel channels. Work is under way at several centers which will enable a machine to prove elementary theorems in certain simple mathematical domains. Such machines may be capable of a modicum of genuine learning. The material outlined in von Neumann's book awaits conversion into mathematics.

CHAPTER 19

GAMOW AND MATHEMATICS: PERSONAL REMINISCENCES

GAMOW'S INTEREST IN MATHEMATICS started in his childhood. In his autobiography, *My World Line,* he mentions the first mathematical problems which directed his curiosity toward the world of numbers through which, by degrees, he became interested in physics. He also describes the courses he took in mathematics and his professors at the University of Leningrad. He says that in those days he was already attracted to the more qualitative more purely conceptual parts of mathematics and, by contrast, he rather disliked the more calculational and formally technical areas of mathematics. So, for example, he was fascinated by the ideas of set theory, the problem of the infinite in general, beyond the purely mathematical treatment of infinity by the infinitesimal calculus in physical problems. He was interested in topology, also in number theory and in combinatorics — the last two being more concerned with manipulation and technical operations *per se* than set theory and topology are.

One can observe this seeming dichotomy in Gamow's choice of interests and work in physics. His overwhelming primary curiosity was directed towards the large lines of the theories which attempt to make us understand the *scheme* of things in the universe — the foundations of physics, the very set up of the dimensions and physical variables, and the constants from which theoretical physics is built — the stage on which all phenomena take place, that is, the nature of space and time in the very small and in the universe at large. Gamow was also concerned with the shape and the history of the cosmos as it has unfolded for us through the work of recent decades, and especially the last few years, as more perplexing spectacles and more mysterious indications of the history in cosmology and cosmogony have emerged. These were his chief interests. At the same time he was fascinated by puzzles, sometimes of special nature, and by very complex devices or "gimmicks" or logical paradoxes. He was co-author of a popular book on *puzzles*.

In this chapter I intend to review Gamow's penchants and inclinations in a variety of mathematical fields and in certain problems of theoretical physics and theoretical astronomy. In innumerable conversations and discussions extending over more than twenty-five years, we touched upon a very great variety of topics which one would be hard put to classify as belonging to either physics or to mathematics proper. I will mention a few of them, hoping to illustrate some of the characteristics of Gamow's insight and of his intuition, both so deep and so far-reaching in different, new directions.

His popular book, *One, Two, Three . . . Infinity — Facts and Speculations of Science*, seems to me to show by the very titles of parts and chapters how he looked upon the role of mathematics in the foundation and language of physics and, more generally, in all natural sciences. I will quote the successive topics in this book and mention in each case some of the flavor of our conversations which took place over the years on these topics and many other, new problems.

Part I is entitled, "Playing with Numbers." It has two chapters, one called "Big Numbers" and the other called "Natural and Artificial Numbers." In this part (the book was destined for the general public) Gamow discusses the systems of numeration, how to define very large numbers, using the powers of 10 and exponential notation, the invention of which he ascribes to Archimedes. He proceeds right away to a popular discussion of infinite sets and powers of infinity with a delightfully simple presentation of George Cantor's ideas of set theory. Gamow was a master of dimensional reasonings. He worked better with exponents than with any other part of arithmetic.

How many discussions and how many hours of arduous thought on a deeper level of knowledge do these topics recall! To the end of his life Gamow was interested in the peculiar dimensionless numbers, e.g., 137 or 1836, the ratio of the masses of protons and electrons, the ratio of the masses of mesons to the mass of proton and neutron, the large number which is the critical Reynold's value and, finally, the enormous numbers expressing the ratios of the gravitational to the electromagnetic forces, numbers of the order of 10^{38} or 10^{40}. For years he was fascinated by the occurrence of 10^{40} and its square in several cosmological interpretations, and especially in the relation between the quantities describing the universe at large and its history, and the constants of the microcosm. Since he always tried to find, even in most abstract theories, motivations or similes, i.e., analogies with precisely understood models, he would try to connect the "large" constants with the age of the universe expressed in elementary time units.

Sometimes, just to tease Gamow, I tried to outdo him in some numerology, and I remember telling him once how the ratio of "the action of the universe" to the quantum of action h of Planck is about $(10^{40})^3$. The action A of the universe, $A = Mc^2T$, we took to be for this purpose the total energy corresponding to its rest mass $E = Mc^2$, where M is about the mass of 10^{78} protons (the number as believed in by Eddington) multiplied by T, the age of the universe T which we took to be about 10^{10} years or 3×10^{17} secs. Indeed, the ratio is about 10^{120} or the cube of Dirac's number. All such observations or speculations Gamow would take half seriously. The more serious discussions, however, concerned the role of the actually or truly infinite in the physical world.

Long ago we agreed that "the problem number one" of astronomy, and indeed of all physical science, should be the question whether the universe is finite or infinite, and we enumerated the various meanings of this question. Gamow was rather inclined to a belief that the Einsteinian universe has a negative, hyperbolic curvature and, therefore, is noncompact, as mathematicians would call it; that the number of stars — or galaxies — would be actually denumerably infinite. When I acquainted him, sometime in the late '30's, with Gödel's result on undecidability in systems of mathematics, we agreed that this question of the finiteness might be undecidable. Apart from the purely temporary difficulties in properly discussing this question, we thought the question could be undecidable in principle. We agreed, though, that for observations like those of the count of galaxies and the indications whether they increase more slowly or faster than the cube of the distance, one would be inclined to *assume* one or the other alternative, "for the sake of elegance" and for drawing conclusions from the finite or infinite model.

When Gamow acquainted me with the pre-einsteinian speculations of Charlier about a hierarchy of condensations — starts, clusters, galaxies, clusters of galaxies, etc., ad infinitum — as a way out of Olbers' paradox, I told him in turn about the mathematical constructions like Cantor's discontinuum and the models of space involving "holes within holes" which could one day play a role in the models of the "elementary particles" space-time geometry. As one of the originators and developers of the "big bang" theory, Gamow did not take very well to steady state theories. I remember how I, myself, tried to make precise the meaning of the uniformity of the universe in the large. If it is actually infinite, then the scale on which to measure this uniformity is something which one should define first. In an actually infinite space, fluctuations can, of course, be arbitrarily large, and the question of operational definition or observed verification of uniformity is not obviously settled. Certainly many of the astronomers seem inclined to the opinion that superclusters of galaxies, and perhaps an infinite sequence of hierarchy, exist — and this is not contradicted by present observations of radio astronomy data.

In the second part of the chapter on natural and artificial numbers, Gamow introduces the imaginary and complex numbers. This brings to mind how I acquainted him with the notion of *p*-adic numbers, their geometry and algebra. In connection with a paper written by Everett and myself on the analogues of the Lorenz transformations in such spaces with p-adic ordinates, Gamow did observe the "outside in" analogy by a "$1/r$ transformation" going into the infinitesimally small with a Charlier type hierarchical universe.

Part II of *One, Two, Three . . . Infinity* is entitled "Space, Time and Einstein." It is characteristic of Gamow's approaches that the discussion starts with combinatorial ideas (on a very popular level, of course), including properties of polyhedra, Euler's formula, etc., and then plunges into topological curiosities or pathologies of "unusual properties of space" connected with its structure as a continuum.

One finds it pleasantly surprising that in this popular booklet both the topological and the algebraical properties of what we call space are so deeply intertwined, and that it is possible to give an account of them on an elementary level. The questions of parity, of mirror symmetries, are introduced there together with discussion of one-sided surfaces — Möbius bands, etc.

Here again let me mention some of the ideas Gamow and I discussed in later years.

It is epistemologically curious to me that the vast and detailed work of mathematicians, which generalizes the notion of space from Euclidean spaces to very abstract constructions, has not been paralleled or followed by a similar effort in possible generalizations of the space-time concepts. The Minkowski space and the Lorenz-Einstein formulation have not been generalized or treated abstractly in analogy to the work which, starting with the examples of concrete three-dimensional spaces, has produced abstract metric spaces or, if one insists on algebraical structure of the space as well, general vector spaces, including the infinitely dimensional ones.

To be sure, Riemannian spaces form a very large class of mathematical objects and the general theory of relativity can be formulated in diverse ways. One could, however, venture an opinion that locally, that is to say topologically in the neighborhood of "points," the classical Riemannian concepts are already too special. The differentiability implies smoothness in

235

the small and Einstein's great program of geometrizing physics may not be easily formulated in such a framework. The reason is, of course, that the phenomena in the small, far from being increasingly "mild" and linear, show the opposite behavior. Even the nature of the departure from the macroscopic changes several times, going to ever smaller dimensions; the forces become more violent and, perhaps, a new substratum for the physical phenomena, i.e., a new notion of space-time structure is called for.

This, then, would seem to involve two possible programs — a search for an alternative topology in the small, and secondly, a very much more general class of mathematical models of "space-time" involving some algebras, perhaps more general than those of presently studied vector spaces. In particular, we discussed the question of whether the "atoms of space" are already endowed with certain algebraical properties of rotation-like character; in other words, whether the spin-like variables reside in the very nature of geometry, or, ultimately, belong to what we call particles of nonlocal character.

The question of whether the present, locally Euclidean, model of space-time is adequate would certainly come next after the problem of the finiteness or infinity of the picture of the universe.

Mathematical analysis in its present state does not allow one to see the consequences of general relativity theory in a form explicit enough for comparison with observations. The mathematical difficulties of solving Einstein's equations are such that a detailed comparison of solutions with observed phenomena is not yet possible. Even questions "in the large" are hard to answer. The topology of a space-time, where one would postulate a Charlier type of distribution of masses (that is to say, an increasing hierarchy of condensations going to infinity), is not immediately clear. At present, answers to such problems in the large generally seem to require a precise knowledge of the solutions of the partial differential equations.

It would be most desirable to be able to make statements about the properties of the topology of a space-time in the large from global properties of distribution of masses, without having to solve *exactly* the partial differential equations. Somehow an algorithm should exist, which would allow one to predict global properties from an imperfect, or perhaps even very inaccurate knowledge of the local behavior of the solutions. Quite generally our feelings were that qualitative questions should have common sense, "up to 10 percent" correct answers by finite algorithms. Should the fundamental or "final" theories involve field equations with prohibitive mathematical complexities, the answer to qualitative or order of magnitude problems would be obtainable by qualitative, order of magnitude arguments and reasonings.

Part III of *One, Two, Three . . . Infinity* is entitled, "Macrocosmos," and the first chapter is called, "The Descending Staircase." In his inimitable way of popularizing difficult ideas, Gamow starts with an explanation of what in recent years has been discussed beautifully by V. Weisskopf in his "quantum ladder" schemas and parables.

If the history of developments in physics in the 19th and 20th centuries is studied to extrapolate the succession of ideas, one certainly discerns an ever-repeating pattern, almost an iteration, of structures. At one time a molecule played the role of the indivisible smallest entity of the physical world, envisaged by the Greek philosophers. This gave place to an assembly of smaller units, the atoms, of which the molecule was a conglomeration. The atom in turn turned out to be representable as a miniature solar system. Its center, the nucleus, in turn is represented as an assembly of nucleons which in turn, as we know now, are a structure. One is entitled to wonder whether the process of building these models will continue indefinitely and whether at some time it may be convenient, and even physically necessary, to assume at least some ways in which the process really goes on ad infinitum — not that the stages repeat exactly or even very similarly, of course.

We often wondered whether the time had come or was coming soon for *"experimenta crucis"* which would in some way indicate whether or not it is essential to imagine finally an indivisible smallest entity. Gamow's strong inclination was to side with proponents of a minimal length λ, of the order of 10^{-13} cm. In fact, in many of his attempts to build a classification of the physical constants based on some three of them, he would choose this λ, along with h and c. He would, however, agree to the possibility, which I often advanced in our conversations, of a discrete *spectrum* of distances going to 0 without assuming continuous values, i.e., a countable infinity of ever smaller quantized spatial dimensions.

At any rate, we know that this "descending staircase" is formed of very heterogeneous steps. For one thing, the very mathematical nature of the successive variables may really differ from one stage to the other. The uncertainty principle and the micromechanics of quantum theory suggest that logically speaking it is perhaps the notion of *sets* and not "*points*" (as evidenced by the real number system) that form the primitive variables. The reality may be stranger yet, since we deal with sets of "undistinguishable" elements.

The next discussion in *One, Two, Three . . . Infinity* involves the role in physics of probability theory and statistical mechanics. In this connection I remember our attempts to establish a little theory for the turbulence of radiation; this in order to see whether, during the first few seconds of the life of the universe (after the big bang occurred), the radiation which contained most of the energy and some of the cosmos could have already shown the first tendency towards forming anisotropies, giving rise to clusters of galaxies and individual galaxies later on. This study, like several other efforts, remains unfinished.

"The Days of Creation," the title of the last chapter, describes in a very popular way the big bang theory and the "expanding horizons." However, the penultimate chapter sandwiched between one on the law of disorder and statistical mechanics, and the last chapter, is entitled "The Riddle of Life." It was some ten years after writing this book that Gamow contributed in a most imaginative and important way to the solution of one of the riddles which life presents. He was the first to suggest, after the discovery by Crick and Watson of the mechanism of replication of the DNA, that the succession of the four letters forming the ladder of the chain form a *code* describing the chemical operation and the activities of the cell.

238

In 1947, when *One, Two, Three . . . Infinity* was first published, one finds in his text only glimpses of these ideas: "Genes as Living Molecules," a paragraph of this next to last chapter, and then what he calls a "semi-fantastic picture" of a long chain of atomic groups, which he calls "a charm bracelet, responsible for the shape, color and other characteristics of a living thing."

A different "ladder," one of increasing organization and combinatorial complexity, seems to lead from a picture of the universe in the large through actions of statistical ensembles of inanimate atoms to the marvelously organized set of chemical games which constitutes life, developed by evolution in a small part of the universe.

CHAPTER 20

MARIAN SMOLUCHOWSKI AND THE THEORY OF PROBABILITIES IN PHYSICS

THE GREAT Polish theoretical physicist, Marian Smoluchowski was a contemporary of Einstein's, older by a few years. He died during the First World War in Cracow in September 1917.

His work in the theory of fluctuations and kinetic theory of gases, especially in the theory of Brownian motions, is well known to physicists. It is interesting though, to survey these pioneering contributions from the perspective of many decades and observe how the ideas of the theory of probabilities influenced the development of the kinetic or "particle" point of view in theoretical physics.

Smoluchowski, at the same time as and independently of Einstein, elaborated and carried forward the ideas of Maxwell and Boltzmann. Some of his fundamental contributions concerned the role of statistical fluctuations in phenomena involving assemblies of particles, and confirmed their importance in explaining phenomena like the Brownian motion and opalescence. One might say that, before him, most studies were concerned with the thermodynamic variables representing the first means or expected values of the relevant random variables. Because it went further into an examination of the deviations and second moments, the work of Smoluchowski gave further confirmation to the reality of a kinetic picture of matter.

It is instructive to consider the increasing role of the probabilistic approach during the years that followed his work. For one thing, quantum theory generalized this approach, extended it, and made it even more basic. It is true that what was called an assembly of elementary particles now forms a more general, if less concrete, picture. Sometimes, in considering nuclear phenomena, one pictures a collection of a considerable number of "virtual" particles or events. The statistical approach like the one, for example, used by Fermi for collision processes in meson production is often very useful even when the number of particles involved is rather small! Here again, the study of fluctuations or deviations from first means is especially important, since the number of the particles is large, though not enormously so. Moreover, in the more classical fields of physics, for instance in hydrodynamics, the ideas which originate historically in thermodynamics (that is statistical treatment) find ever wider applications — to mention only the theory of turbulence in this connection.

It is interesting, even today, to reread some of Smoluchowski's general speculations on the idea of chance and the origin of principles of the theory of probability in physics and then follow the subsequent development of ideas in this direction.

A *"Selection of Philosophical Writings of Smoluchowski"* published in Poland in 1955 contains a number of his more general and celebrated articles like the one on the second law of thermodynamics, or a beautiful one on *"The Subject, Problems, and Methods in Physics."* It also contains several of his lectures. The one on the importance of exact sciences in general education full of relevant remarks on the relative role of traditional classical education versus a more technical one is still a vital topic.

One gets a feeling of the scientific atmosphere and life in European universities at the turn of the last century by following the travels of Smoluchowski in Germany, Italy, France, and England and his work in Vienna, Lwów, and Cracow. It is interesting and perhaps melancholy to compare the degree of mutual influence of different centers then, with the degree of scientific cooperation between central and eastern Europe on one hand and the West on the other, at the present time.

Smoluchowski's studies and his post-doctoral travels give a very vivid impression of the genetic development of many of the ideas of modern physics in their germinating stages. In physics, this period was, perhaps, one of a calm before the storm of the full impact of the theory of relativity and quantum theory. One feels inklings of an impending revolution in the foundations of physics, and since the present time, perhaps, gives a similar impression, it may be instructive to follow in retrospect the appearance at that time of the signs and omens of change to come.

Smoluchowski was born in Vienna in 1872, the son of a Cracow lawyer who was later secretary of the Chancellery at the Court of the Emperor Francis Joseph. Young Smoluchowski had the benefit of exceptionally favorable conditions in his childhood and youth. His mother, very well educated, created at home an artistic and intellectual atmosphere. Vienna, the capital, was a brilliant center of art and science in its day. Smoluchowski studied at the famous Theresianum, the best "gymnasium," attended by sons of the nobility and high government officials.

In his high school years, young Smoluchowski showed interest in science, especially astronomy, and, thanks to his teacher, became interested in problems of physics. After graduating from the gymnasium, he entered the University of Vienna where he elected physics and mathematics as his fields and attended the lectures of Boltzmann. In all his work he kept a predilection for stressing equally theoretical and experimental aspects. His admiration for Boltzmann continued throughout the rest of his life.

He published his first scientific paper during the second year of his university studies. He graduated with the highest honors and received his doctorate *"Sub Auspiciis Imperatoris."* (This might amuse the modern reader: the most brilliant graduates at Austrian universities were distinguished by a special reward given in the name of the Emperor, including a diamond ring.)

Smoluchowski spent the year 1895-1896 in Paris working with Lipmann. He expressed in his letters a great appreciation of the intellectual atmosphere of Paris but also some disappointment at the lack of contact between the professors and the younger scientists.

The following year, before going to Glasgow, he spent the summer mountaineering in Scotland. He had a real passion for this activity since his childhood and continued throughout his life.

The sojourn in Glasgow made a very strong impression on young Smoluchowski, and the influence was visible in his later life; many of his future attitudes and ideas had their source there, some of them due to Lord Kelvin. Smoluchowski was greatly impressed by the spirit of the laboratory to which visitors came from all over the world. For years, Smoluchowski remembered the high level of experimental and theoretical work there, in seeming contradiction to the very modest buildings and inadequate facilities. In a letter to his brother dated January, 1897, he wrote: "Lord Kelvin seems always greatly excited. Every day some different problem seems most interesting to him and he forgets everything else. He is impatient and cannot perform the experiments himself but could keep busy with his ideas a whole army of experimental physicists."

The next year, after a few months in Berlin he returned to Vienna and was appointed Dozent (lecturer), and lectured in mathematics and quantum theoretical physics. He was offered a professorship in Bombay, but after considerable hesitation, he decided not to accept it, and instead, in 1899, became a lecturer in Lwów. The next year, at 28, he was promoted to a professorship of theoretical physics. This made him the youngest professor in Austria.

In the Spring of 1901, he visited Glasgow for the 450th anniversary of the foundation of the university and received an honorary doctorate there. His marriage also took place in 1901.

The fourteen years of his stay in Lwów formed a period of increasing activity and ever-expanding theoretical work. His teaching was extremely successful, and inspired a whole generation of young physicists and mathematicians. Despite his popularity with students and colleagues, however, he considered teaching to be a secondary aim. He continued his research and worked feverishly at publishing a whole series of important papers. The scientific atmosphere at the university was, before his coming, rather poor, at least in his field; it was mainly through correspondence that Smoluchowski kept contact with other physicists of the world.

From August 1905, to April 1906, taking a leave of absence, he went to Cambridge and worked at the famous Cavendish Laboratory. Among the people in his own fields of interest with whom Smoluchowski had especially close contact one should name the mathematicians Ball and Hobson and the physicists J.J. Thompson and Rutherford.

Upon returning to Lwów, full of new ideas, he started a series of papers, simultaneously with and independently of Einstein, and in rather more detail, on the kinetic theory of gases, and developed the theory of Brownian motion.

Smoluchowski left a series of notebooks full of calculations and remarks especially noting the analogies between problems belonging to different parts of physics. The perception of analogy often precedes the discovery of more general principles. (This intuitive faculty is of transcendental value even in pure mathematics: the writer recalls a remark by S. Banach, a great Polish mathematician: "Good mathematicians notice the analogies between theories and between methods of proof. The very great ones see the analogies between analogies.")

Smoluchowski was very adept in designing and building instruments. He happened to be exceedingly good at glass blowing, better than the professional mechanics at his institute. To illustrate anecdotally his liking for simple experimentation, the writer was told by Mrs. Smoluchowska how, one morning he asked her for a flat kitchen vessel and she handed him a salad bowl. He told her that he was trying to imitate a possible mechanism for the appearance of mountains on the surface of the earth. A few days later, one saw a layer of gelatin covered by a surface of mercury folding into mountain chains and ridges. This was the first step on experiments which gave rise to series of papers, on a theory of mountain formation.

The year brought a series of experiments confirming Lord Raleigh's theory that the blue of the sky was caused by the scattering of light.

245

As the most eminent scientist at the University of Lwów, he received many invitations to public office. For example, he was invited, in succession, to be a member of the City Council, Parliament, and Council of State. He refused all of these and did not even accept an offer of the presidency of the Lwów Scientific Society, realizing that administrative work would take too much of his time. Nevertheless he was an active member of the Polish Physical Society whose first President he became. He became Dean of the Faculty of Philosophy (Arts and Sciences in our universities) during a politically difficult period for the Polish part of Austria.

There was, at that time, no textbook of theoretical physics in the Polish language. Students have edited a mimeographed collection of Smoluchowski's lectures. These he planned to use sometime in the future to prepare a large textbook of theoretical physics. As a member of the Academy of Sciences, he gave several public lectures.

In 1913, he was invited to Göttingen to give lectures on the kinetic theory of matter; the other lecturers in the series were H. G. Lorentz, M. Planck, A. Sommerfeld, and P. Debye. That year he was offered the chair of experimental physics at Cracow. Even though the University at Lwów was loath to let him leave, Smoluchowski decided to move to the University of Cracow, which, founded early in the 14th century, was the oldest university in Poland, and which offered a new physics institute and the chance of being closer to the centers of western Europe.

World War I started the following year. Smoluchowski was evacuated with many others to Vienna during the first few months of the war. He was called into the army and worked in military censorship.

In September 1916 he was invited to give lectures on the Brownian theory of motion and coagulation in Göttingen. In 1917 he published his famous paper on the mathematical theory of kinetics and colloidal suspensions. This made him one of the founders of the theory.

In 1916 and 1917, as a Dean of the Faculty of Philosophy, he had to make trips to Vienna to get funds for the university and nominations confirmed. Also, as a member of the Organizational Committee of the newly formed School of Mines in Cracow, he made trips to Germany and Austria. All these activities were exhausting and only with difficulty, did he find free hours for scientific work. Mountaineering, skiing, and music became luxuries which were no longer possible, and he dreamed of being able to return to such relaxations "after the war." The horrors of war depressed him greatly, and he complained that civilization was perishing.

Political frictions at the university and among the Polish population in general affected him, and he often became pessimistic. It was in such an atmosphere that he received an invitation to a chair of theoretical physics in Vienna. The University of Cracow kept him by making him rector (President).

Smoluchowski looked forward to his inaugural address as president. The title "On the Uniformity of the Laws of Nature" was indicative of his interests during the last years of his life. They became more general and centered in an attempted synthesis of phenomena.

He died suddenly from dysentery after a very brief illness on September 5, 1917, at the age of 45, without learning that the Academy of Sciences at Göttingen had elected him a member.

This brief account of external circumstances and personal contacts shows how political turmoil can influence scientific activity.

The present-day reader accustomed to the great degree of organization and collaboration of modern scientific work will note that Smoluchowski worked as a young professor outside the great centers of scientific activity of the time.

It is interesting to see how it was possible for a person of his exceptionally high ability, to get to the forefront of European thought in physics, even though the milieu in which he worked as a young professor was relatively isolated and without tradition in science. Nevertheless it was possible to start pioneering work in the relatively new field of statistical mechanics and get to the forefront of world science once a catalyzing contact with other minds had been made.

A similar situation arose in Poland just after the end of World War I: A small group of mathematicians managed to establish by their enthusiastic work, an impressive new school of mathematics in fields such as set theory, topology, foundations of mathematics, despite the lack of a tradition of work in mathematics in that country.

Such precedents might be of note should one be surprised at the rapid rise of significant work in experimental and theoretical physics in eastern Europe at the present time.

The probability theory has played an expanding role in physics since the premature death of Smoluchowski. He did much to start these applications and anticipate their progress.

One of his mathematical achievements was the clarification of the role of the so-called ergodic hypothesis of Boltzmann. In a dynamical system composed of N particles, the development in time is purely deterministic. The positions and velocities are in principle calculable for all time. The whole system can be presented as one point in a space of $6N$ dimensions. This point will move as time goes on through certain of the positions within the available phase-space corresponding to the given total energy of the system. That is to say, the representative point will describe a line on a surface of $6N - 1$ dimensions.

Boltzmann conjectured that this line will ultimately go through every point of this surface. It was noticed, among others, by Rosenthal that this is topologically impossible. A weaker statement, however, would have been sufficient to imply the conclusions of Boltzmann, namely, the assumption that the point will get arbitrarily close to any point on the surface and will move in such a way that the time of sojourn in any sub-region of this surface will be, asymptotically, proportional to the volume of the sub-region. In this sense, the motion will present certain features of a *random* sequence of points where one would expect, if one selected positions at random rather than by deterministic equations, that these points will cover the space uniformly densely.

This postulate of ergodicity is indispensable for the rigorous foundations of statistical mechanics. This property still remains unproved for most of the actual dynamical systems such as for example the *n*-body problem. General theorems, however, were proved by J. von Neumann and G. D. Birkhoff, asserting the existence of such averages in time of sojourn in sub-regions for "almost every" point of the phase-space. This is the celebrated ergodic theorem. Furthermore, one can also prove that for quite general, volume preserving, flows of phase-spaces, such sojourn times are for "almost all points" indeed necessarily proportional to the volumes of the sub-regions in question.

One of the points of debate which was so crucial during Smoluchowski's time was thus settled. It shows that deterministic theories of phenomena, *in general,* lead to *randomlike* sequences of points in the representative space. Let us quote a sentence from the posthumous paper of Smoluchowski in *Naturwissenschaften* (1918). "It seems to us that it is very important even for the philosopher that one can prove, even though only in a limited part of mathematical physics, that the idea of probability in the ordinary sense of a regular frequency of random effects has also an objective meaning, namely, that the idea and genesis of randomness can be made rigorously precise also if one rigorously follows the determinism; the law of large numbers comes then not as a mystical principle and not as a purely empirical fact but as a simple mathematical result of the special form which the law of causality determines in such cases." This prophetic sentence certainly is justified by the ergodic theorems proved for the first time only in 1931.

It is true that the work of mathematicians—even in the classical part of statistical mechanics, that is to say not in the quantum theoretical formulations, establishes a foundation merely for the beginning parts of thermodynamics. An analogous and rigorous treatment of the so-called H-theorem and the Boltzmann equations is not yet completed. It is interesting that Smoluchowski realized the difference in the logical structure of the theories of Maxwell, Boltzmann, on one hand, and the statistical mechanics of Gibbs. Again quoting from the same article, he says explicitly that "the difference between the two very likely consists of the fact that the former is based on certain ideas from the theory of probability, extremely plausible for these physical systems but not rigorously proved, whereas the latter avoiding these ideas, is based entirely on postulated statistical properties." This feeling for the logical structure of physical theories is extremely noteworthy when one remembers the time when these remarks were written, that is before the great vogue of axiomatization and the development of the foundations of mathematics.

The statistical or probabilistic points of view are now used more and more widely even in the domain of classical physics. The analogies of the ergodic theorems which rigorized such treatment of systems composed of a finite number of particles came to the forefront of the work of recent years in the study of motions of continua. We have in mind statistical theories of turbulence in gases or liquids which involve infinitely many — a continuum of points. Also, the new field of magnetohydrodynamics, in the classical formulation at least, has infinitely many degrees of freedom. The equivalent of the work of Maxwell, Boltzmann, Gibbs, Smoluchowski, Poincaré, and Birkhoff, presents here a much vaster and mathematically more difficult task.

One might, at first, form the impression that the introduction of infinitely many particles or degrees of freedom is really not basic and can be attributed only to the fact that calculus uses such idealizations convenient mathematically for algorithms of analysis but not fundamentally necessary since gases and liquids are composed of finite assemblies of atoms. In this case, the remarks of the last few sentences would have only a methodological relevance. Unfortunately, the situation is much more difficult. The ideas of quantum theory and the role of field theory underlying the quantum theoretical description mean that one has not so far, got rid of the true continuum as a basic notion in physics. In the beginning of the work of Maxwell *et al.*, and for the purpose of explaining the phenomena which they set themselves to do, it was sufficient to think of atoms as spheres or even as points which collided with or deflected each other, without the necessity of considering their "structure" which influences such events.

How would Smoluchowski feel, one wonders, toward the present day formulations of quantum theory which deal *ab initio* with probabilistic concepts? One might say, of course, that the theory itself is as deterministic, or logical as ever, but the "primitive term" of it is not a point in ordinary space but rather a probability distribution. Would he be on Einstein's side in the discussions with Bohr regarding the real meaning of determinism in physical theory? Smoluchowski discussed in his papers the necessity for continuous introduction of "hidden parameters."

In one of his papers on the uniformity of the laws of nature, he asks the question, "Why is it that the laws of nature seem as simple as they are? It is undoubtedly because we look only at small sections of phenomena or only at special features of the physical world. Now any analytic function is smooth or linear in the small." This paraphrased statement, reminiscent of Poincaré, also bears philosophical affinity to the beliefs which Einstein himself had until the end of his life. If one hopes to geometrize physics in the way that Einstein wanted to, that is to say, to describe field quantities by differentiable functions, one states a belief in the ultimate simplicity of phenomena in the small. The experience of the last decades, with the ever more perplexing variety of forces and objects in the very small makes such hopes appear, at least for the present time, rather remote.

During Smoluchowski's lifetime, W. Ostwald defined and discussed two sharply distinct types of creative scientists: the classicists and the romantics. If one accepts such division, Smoluchowski was certainly a representative of the latter class. Together with his fine mathematical feeling and technique, he constantly searched for the transcendental relative to the then accepted schemata of physics and searched for the hidden parameters, forces, as one would say today, operators, underlying the levels of physical description of the day.

251

CHAPTER 21

KAZIMIERZ KURATOWSKI

KAZIMIERZ KURATOWSKI died in Warsaw, June 18, 1980 at the age of 84. A great mathematician, one of the creators of modern topology, he had an enormous influence in research and in education, not only in Poland but throughout the world. One of the founders of the famous Polish mathematics periodical *Fundamenta Mathematicae,* he was responsible for the development of the new spirit of modern mathematics. After the end of World War II his great merit was the reestablishment and reorganization of numerous mathematical activities in Poland. His influence also spread to the United States through his many students.

In what follows I shall try to reminisce about my early contacts with him before World War II in Poland, and later, and will attempt to sketch some traits of his personality and his activities.

I was his first doctoral student and received my Doctorate of Science at the Polytechnic Institute in Lwów in the General Faculty in 1933. Throughout high school and partly thanks to a teacher, Professor Zawirski, who was also a Dozent at the University, I was interested in mathematics. In the fall of 1927 I entered the Polytechnic Institute in the division of electrical engineering and as a freshman took a course in set theory (a general abstract foundation of mathematics) taught by a new young professor freshly arrived from Warsaw, named Kuratowski. At that time he was slim, rather short in stature, lively in speech but slow and measured in his movements, a quality he retained throughout his life. The course was attended by some fifteen students and the lectures were, in contrast to my high school experiences, monuments of logic, clarity, systematic presentation, and preparation.

That same year, just before the Christmas recess, Kuratowski mentioned a still unresolved problem about a property of transformations in set theory. I thought for hours about this problem, and at the end of a week believed I had found a solution. When classes resumed I communicated it to Kuratowski with great excitement. It proved to be correct!

From that time on I visited him in his office every morning after classes, amazed at his accessibility and at the obvious interest he took in me. I began to think about switching from engineering to mathematics, but the decision was a hard one. There were few, if any, university positions vacant for graduates in mathematics, and my family was initially doubtful about the practical outlook of undertaking such studies in the General Faculty. So I postponed the decision until, I told myself, I could solve another open problem of mathematics. In the spring of 1928 I was fortunate enough to succeed in solving a second problem which had come up in Kuratowski's course, and he decided that these two results should be published as two papers in *Fundamenta*.

I should add here that Kuratowski was a bachelor at the time and lived in a *pension* where he returned for lunch after classes. I used to accompany him home and we walked the mile or so between the Institute and his domicile conversing almost exclusively about mathematics. These conversations are still vivid in my mind, and it is only many years later that I have come to appreciate the patience and the interest this thirty-one-year-old professor was showing in an eighteen-year old very eager student. After lunch at home with my parents, I frequently returned to the Institute not only for other classes, but for more discussions with some of the other professors of mathematics such as Banach, and some of the younger ones like Mazur and Auerbach who also became my friends. At the time Banach was a professor at the University but he occasionally taught additional courses at the Polytechnic Institute.

In 1929 Kuratowski married. I remember that he sometimes took a plane, a rare occurrence in those days, from Lwów to Warsaw to visit his fiancée. He and his wife later settled in Lwów and instead of our daily walks, I remember being invited to their house once or twice a week for mathematical discussions.

In 1930 I managed to solve another problem which Kuratowski had mentioned. It was a problem of his and Banach's concerning possible generalization and strengthening of their joint result.

During these years of studies I was very remiss in writing up original work and, worse yet, could not bring myself to take the exams in the regular courses. It was at Kuratowski's insistence that my early papers were written up and published, and thanks to this original work I was, exceptionally, allowed to take the examinations all at once. In 1932, I received my M.A. and in 1933 obtained my doctorate, under Kuratowski.

I should note here that the Lwów mathematicians had the habit of gathering in coffee houses and tea-rooms. Banach, Stozek and others spent hours between lectures sipping coffee and discussing mathematics at the Café Roma, then at the Szkocka, almost daily. Steinhaus and Kuratowski on the whole preferred the more genteel atmosphere of Zaleski's tea room some two hundred yards away.

The Mathematical Society met at the University almost every Saturday evening and invariably some people repaired afterward to the coffee houses to continue the discussions. Kuratowski attracted to Lwów mathematicians from Warsaw, Sierpinski, Mazurkiewicz, Tarski, and Knaster paid us frequent visits. Karol Borsuk, who was just a few years older than I, came to stay for several months to work with him and also with me. We became close friends and collaborators from that time on. I am mentioning all this to indicate how active and widespread the mathematical life was in Lwów.

Shortly after I received my doctorate, Kuratowski abruptly left Lwów and returned to Warsaw to become a professor at the University. A chair was especially created for him by the then minister of education W. Jędrzejewicz, now living in the

United States. The reason for his departure was this: The General Faculty of the Polytechnic Institute was abolished by decree from Warsaw, I suppose for economic reasons. It had awarded only two doctoral degrees, mine and Jan Blaton's, a young physicist who worked under Rubinowicz. Kuratowski and several other professors lost their positions. (Among them, Jan de Rosen, a famous painter and architect who later came to the United States and created the celebrated mosaics of the Washington D.C. cathedral.)

I came to the United States in December 1935 on an invitation from John von Neumann for a few months' visit to the Institute for Advanced Study in Princeton. Great was my surprise when, the following March, Kuratowski appeared at the Institute while on a lecture tour of American universities. This was his first visit to the United States. There were many more to come after World War II. While he was in Princeton I received an offer of a position at Harvard. His testimonial on my behalf may have played a role. When I asked him whether I ought to accept the position, he encouraged me to do so in view of the scarcity of openings in Poland.

Every summer, including 1939, I returned to Poland to visit my family and friends, and continued to collaborate actively with the Polish mathematicians. Kuratowski and Borsuk invited me several times to stop off in Warsaw on the way to Lwów and give some lectures at the University. He asked me once to visit him at his summer villa outside Warsaw. To my great surprise I found him playing a fair game of tennis. It had not occurred to me that he could be interested in things other than mathematics!

Two weeks after I had returned to Harvard in 1939 the German invasion of Poland took place and all contacts with Kuratowski and other Polish colleagues ceased.

I learned in a round-about way of the tragedies that befell the Lwów professors. Stozek, Lomnicki, Bartel, and many others were gathered by the Nazis shortly after their entry into Lwów and executed, some with members of their families. Kuratowski was in Warsaw when the Germans occupied the city. I had no news about him or any of the other Warsaw mathematicians until after the war. Only then did I learn of the horrors they lived through.

Kuratowski's "Notes for an Autobiography" appeared in *Kultura* in 1979. After his death, his daughter, Sofja Kuratowska, a medical doctor and research biologist, found on his desk a manuscript of reminiscences. Fragments were published in *Kultura*. They are supposed to appear soon in book form in Poland.

It is impossible for me to give even only a bare idea of the moving account of his Odyssey during the occupation: his narrow escapes from arrest by the Gestapo, the gruesome happenings which befell his friends and colleagues, all the while his being very active in the underground work of the secret university. His equanimity and courage come through every page. Perhaps this helped him escape the terrible fate shared by so many others.

His detailed account continues with the period after the liberation. It describes his decisive, really Herculean work to reestablish not only mathematics but other sciences as well in postwar Poland. He was named director of the Mathematics Institute of the Polish Academy of Sciences. As he said, this became the very core of his life.

It is heartbreaking to read his account of the reopening of the academic year 1945/46 amid the ruins of the city which, he wrote, "was slowly being reborn as a Phoenix from the ashes."

I cannot describe, even in general terms, his titanic organizational and educational efforts. It seems also almost superhuman that he could have continued them alongside original creative scientific work.

His creative mathematical research continued unabated to the end of his life. There now exists an English translation of the two volumes of his monumental book on topology, which is a veritable bible of the foundations of the subject. In another book, *Half a Century of Polish Mathematics*, he describes the development of the Polish School and his role in both its subject matter and its organization and reorganization. How mathematical logic, set theory, topology, functional analysis, and other branches of mathematics flourished between 1918 and 1940, and the rebirth of the School after World War II.

Kuratowski received many academic honors including membership in numerous scientific academies. He was very prominent and active in the International Mathematical Union and was, *inter alia* a member of the committee which awarded the Balsan Prize. One of its first recipients was Pope John XXIII.

Our postwar contacts resumed, first by correspondence, then in person during his frequent visits to American universities and my infrequent ones to Poland. In the late forties he was able to stop off in New Mexico and my wife and I drove him to El Paso on his way to other university centers. By then he had become somewhat portly and even slower in his movements than I remembered. My wife was very struck by his measured and smiling ways and his absolute imperturbability when our car broke down in the middle of the desert and it took several hours before we could obtain help. A similar episode happened when we visited him in Poland. He invited us to drive with him to Cracow but his car developed troubles on the outskirts of Warsaw. He remained patiently unruffled while his chauffeur frantically tried to get it repaired. After his war experiences, he said, nothing could ever upset him anymore.

As a host he was unequalled in graciousness and savoir faire. During a brief morning visit at the Mathematics Institute in Warsaw he received us in his large office. We sat with him in deep leather upholstered chairs in a comfortable conversation corner while, in addition to the inevitable tea, a uniformed attendant appeared with mounds of wild strawberries with whipped cream.

As the first of his many students, I bear a great debt of gratitude for his decisive influence on me during my early youth, for the initial choice of my career, and for his having introduced me to the world of mathematics and mathematicians. There are many in America today who have been his students and his students' students. His unceasing work and steadfast character have been examples to follow. His mind was remarkable for its clarity, taste, moderation and common sense approach to the role of abstraction in the great and varied fields of mathematics.

CHAPTER 22

STEFAN BANACH

STEFAN BANACH died in Europe shortly after the end of World War II. The great interest aroused in this country by his work is well known. In fact, in one of Banach's main fields of work, the theory of linear spaces of infinitely many dimensions, the American school has developed and continues to contribute very important results. It was a rather amazing coincidence of scientific intuition which focused the work of many mathematicians, Polish and American, on this same field, a field which grew so richly in the period between the two world wars. Actually, it was E. H. Moore who was the precursor of the doctrine of applying abstract algebraic methods to many concrete problems of analysis. It is not necessary to describe the achievements of Banach and his students which center around his monograph on the theory of linear operations.

Banach's work brought out for the first time in the *general* case the success of the methods of geometric and algebraic approach to problems in linear analysis—far beyond the more formal discoveries of Volterra, Hadamard, and their successors. His results embraced more general spaces than the work of such mathematicians as Hilbert, Schmidt, Riesz, von Neumann, Stone, and others. Many mathematicians, especially the younger ones in the United States, took up this idea of geometric and algebraic study of linear function spaces, and the work is still going on vigorously and producing rather important results.

Also, it is not necessary to stress the possibilities of the application of similar methods of approach to problems of non-linear analysis, at present so numerous and important for immediate applications. Banach had a number of results (unfortunately unpublished) on the theory of polynomials and analytical operators in a class of suitably defined infinitely-dimensional spaces.

He had in preparation a sequel to his volume on linear operators dealing with spaces on which one could study non-linear transformations. Various approaches to the study of "non-linear analysis" have of course been made before by mathematicians. In the opinion of the writer the proper definition of a natural class of spaces on which these more general operators should be studied has not yet been found. Very often in the history of mathematical theories one of the factors deciding the success of a new discipline is the proper amount of generality for the class of objects under discussion. It should contain a large class of examples already existing in mathematics and not be overburdened by too many arbitrary constructions.

Banach's contributions to fields of mathematics grouped around the theory of functions of real variables, set theory, and general groups are equally important and include some of the most elegant and final results in these fields. To give only a few examples, the specialist will remember his theorems on functions of bounded variation, the theorems on differentiability of functions, and his paper on the length of curves and the area of surfaces. His work of the early twenties was characterized by extreme elegance and perspicacity of proofs. One remembers the originality of his proof of Vitali's decomposition theorem and the beautiful set-theoretical investigations on one-to-one transformations of abstract sets. His proof of the theorem stating that a set which is locally of first category is of first category in the large, even for the non-separable case, is an example of his ingenuity. Of course, most mathematicians know Banach's results on measure in Euclidean spaces and in abstract sets (containing the results obtained jointly with Tarski on the paradoxical decomposition of two spheres of different radii into an equal number of pairwise congruent sets). Banach's theories of generalized limits for sequences of numbers and of functions, a joint work with Mazur, find a growing number of applications. He wrote also two interesting papers on general metric groups.

When we survey his work, the role played by the set theoretical and topological methods in more classical mathematical disciplines becomes obvious. For instance, the very important work of Schauder (a student of Banach who was killed during the war) in which he establishes existence theorems in the theory of differential equations uses the method of general function spaces. As other examples of the success of these methods of analysis one has only to point out Banach's work in the theory of general orthogonal series or general integrals.

Among Banach's writings, in addition to his well known monograph on linear operations, are a university textbook on differential and integral calculus in two volumes and a university text on mechanics — the latter, in the opinion of the writer, a masterful presentation, extremely worth translating into English.

Banach was born in 1892. His family was poor and he had very little conventional schooling at first. Thus he was mostly self-educated when he went to the Polytechnic Institute in Lwów. It is said that it was Steinhaus who accidentally discovered Banach's talent by overhearing a mathematical conversation between Banach and another student on a park bench in Cracow. Banach and Steinhaus were to become later the closest collaborators and the senior members of the Lwów school of mathematics. Banach was a student at the Polytechnic Institute between 1910 and 1914. He obtained his Ph.D. in 1920, and was instructor at the Polytechnic Institute between 1920 and 1922, became a dozent and an "extraordinary professor" at the Lwów University in 1922, and a professor in 1929. He was a member of the Polish Academy of Sciences and won several prizes including the prize of the Academy and a scientific prize of the City of Lwów. The writer has no precise knowledge of his life and work from the outbreak of the war to his premature death in the fall of 1945.

Banach worked in periods of great intensity separated by stretches of apparent inactivity. During the latter, however, his mind kept working on the selection of statements that would best serve as focal theorems in the next field of study. He liked constant mathematical discussion with friends and students. The writer remembers a mathematical session with Mazur and Banach lasting seventeen hours without interruption except for meals. In general both the Lwów and Warsaw mathematical schools were fond of and successful in collaboration. A large proportion of the papers published were written by more than one author. It was Banach and, in Warsaw, Sierpiński who were mainly responsible for this development.

Much of the mathematical work was carried on in a way not usual in American mathematical centers. The mathematicians in Lwów met not only in their classrooms and offices but spent long hours every day in two coffee houses which served as informal meeting places. They discussed problems over coffee or beer, and marble table tops and napkins took the place of blackboards. It was hard to outlast or outdrink Banach during these sessions. He was always there. The weekly (Saturday night!) meetings of the mathematical society, where papers were presented, provided the more formal discussions.

Banach's facility for proposing problems illuminating entire sections of mathematical disciplines was very great. These, properly formulated, can often play the role of *experimenta crucis* in a physical theory. The constant challenge of these questions, proposed by mathematicians of the group over which Banach presided, stimulated the gradual progress of the theories, now so well known, in functional analysis and real variables. He took up eagerly the challenge of any important question connected with the domain in which he was working at the time, and he could spend months on it.

Banach's publications reflect only a part of his mathematical powers. The diversity of his interests in mathematics surpasses by far that shown in his papers. His personal influence on other mathematicians in Lwów, and in Poland, was in many cases decisive. It stands out as one of the main traits of this remarkable period of twenty years when so much mathematical work was accomplished and so many high hopes were raised — all to be interrupted by World War II.

CHAPTER 23

A CONCLUDING PAEAN

THE ONE HUNDREDTH ANNIVERSARY of Einstein's birth marked the end of a century-long period during which the achievements of mankind in the sciences and in the understanding and control of the forces of nature made more progress, quantitatively and perhaps as much in qualitative understanding, as in all previous history. Einstein's ideas contributed to this in a very large measure, not only in theoretical physics but also in its applications to technology.

The concepts of special relativity appeared 'strange' in the context of the physics of his day. The general theory of relativity introduced a lien between the notions of geometry and those of physical forces. His work on Brownian motion and on the emission and absorption of photons brought in a fundamental connection between the world of the probabilistic basis of phenomena and the quantum theoretical class of ideas.

Since his work in the early twentieth century, the 'strangeness' of the astronomical, the physical and even the mathematical worlds became increasingly manifest. The recent discoveries of stars like pulsars, the mysterious objects like quasars — galaxies perhaps — with their enormous output of energy in explosions — were certainly totally unexpected in the picture of the cosmos during the first half of this century. The same 'strangeness' seems to pervade the world of the infinitesimally small with its plethora of subnuclear particles and perhaps infinity of stages of real of virtual arrangement of matter. In mathematics itself, the belief

263

held during the nineteenth and the first third of this century of the possibility of a complete axiomatic basis for all of logic and mathematics, so succinctly expressed by Hilbert, was shattered by Gödel's results. Recent developments show the possibility of 'non-Cantorian' models of infinity in set theory and even in analysis. The strange large cardinals in one hand, and the incompleteness of the mathematics encompassed by any computing algorithms on machines on the other, present a perplexing situation. In this sense the foundations of logic show the demise of the 'absolute' and the 'relative' way of thinking now evident in mathematics.

It seems that Einstein's greatest forte was his qualitative mold of thought. The technical executions of such ideas present often wonderful formulations, but they were subject later to enlargement and change. Thus, for instance, von Neumann's work on foundations of mathematics, and his mathematical basis of quantum theory preceed the subsequent developments.

The age of electronic computers, of lasers, and of nuclear energy has its origins in the study of the behavior of formal systems on one hand, and on Einstein's work on stimulated emissions and on the equivalence of mass and energy on the other.

The limitations of the formal system are now apparent. The possible infinity of the scale of physical phenomena seem analogous to the ancient proofs of *impossibility* — the irrationality of $\sqrt{2}$, the non-constructibility of $\sqrt[3]{2}$, the infinity of prime numbers. The *impossibility* of sending signals to points separated by the light cone is in this vein.

In addition to the revolutionary developments in mathematical logic and foundations of mathematics, enormous progress was made in such 'very pure' mathematical disciplines as number theory, algebraic geometry, combinatorial and geometrical topologies and combinatorics itself, during these one hundred years. I hope that I have conveyed some of the wonders of these developments in this volume.